U.S. NATIONAL FORESTS IN THE WEST

*See end pages for map of
U.S. National Forests in the East*

s and Clark

MONTANA

NORTH DAKOTA

a

Gallatin

Custer   Custer   Custer

SOUTH DAKOTA

dger-Teton   Big Horn

Black Hills

Shoshone

WYOMING   Medicine Bow

Nebraska   Nebraska

Nebraska

NEBRASKA

White
River   Routt   COLORADO

Ashley   Grand
Mesa   Roosevelt

LaSal   Arapaho

ncompahgre   Pike

LaSal   Gunnison

San Isabel   KANSAS

aSal   San Juan   Rio
Grande

Carson   Carson

Santa Fe   Santa Fe

Cibola   OKLAHOMA

greaves   Cibola

Cibola

Gila   Lincoln

NEW MEXICO   TEXAS

"This guide book says welcome."

"It has been said that the national forests of the U.S. are a well-kept secret. With Robert Mohlenbrock's book, it is safe to say the secret is out. The recreationist can share with him the special botanical, geological, and scenic attractions of the 153 national forests he has visited across the country by car, by foot, and even by canoe. These outdoor wonders belong to all of us, and this guide book says welcome."

—R. MAX PETERSON,
Chief, United States Forest Service

Also by Robert H. Mohlenbrock

Where Have All the Wildflowers Gone?
Forest Trees of Illinois
The Flora of Illinois (10 volumes)
Wildflowers of the United States, Volume I: Wildflowers
of the Northeastern States (co-author)

THE FIELD GUIDE TO U.S. NATIONAL FORESTS

Robert H. Mohlenbrock

CONGDON & WEED, INC.
NEW YORK

Copyright © 1984 by Robert H. Mohlenbrock

Library of Congress Cataloging in Publication Data

Mohlenbrock, Robert H., 1931–
The field guide to U.S. national forests.

Includes index.
1. Forest reserves—United States—Guide-books.
I. Title. II. Title: Field guide to US national forests.
SD426.M64   1984   917.3′04927   83-25218
ISBN 0-86553-117-X
ISBN 0-312-92206-X (St. Martin's Press)

Published by Congdon & Weed, Inc.
298 Fifth Avenue, New York, N.Y. 10001
Distributed by St. Martin's Press
175 Fifth Avenue, New York, N.Y. 10010
Published simultaneously in Canada by Methuen Publications
2330 Midland Avenue, Agincourt, Ontario M1S 1P7

Design by Irving Perkins
Drawings by Pamela Congdon
Cover design by Regine de Toledo

COVER ILLUSTRATIONS
Front cover (left to right, top to bottom):
Brilliant Vermilion Cliffs, Dixie National Forest, Utah
Bighorn Ram, Lolo National Forest, Montana
Warner Lake and Haystack Mountain, Manti-LaSal National Forest, Utah
Quinault Rain Forest, Olympic National Forest, Washington
Fire Pink, Nantahala National Forest, North Carolina
Back cover (left to right, top to bottom):
Rocky Mountain Goat, Lewis and Clark National Forest, Montana
Stream near Silver Mine, Mark Twain National Forest, Missouri
Spectacular Multnomah Falls, Mt. Hood National Forest, Oregon
Natural Rock Bridge, Daniel Boone National Forest, Kentucky
Cape Perpetua, Siskiyou National Forest, Oregon
Raccoon, Daniel Boone National Forest, Kentucky

Credit for the photographs of the Bighorn Ram, Rocky Mountain Goat,
and Raccoon goes to wildlife photographer Tom J. Ulrich. All other
photographs were taken by the author.

*To all my graduate students
past and present
who have shared with me a love
for our natural heritage*

PREFACE

Several years ago, after having visited most of the national parks of the United States and finding them increasingly more congested, I started taking working vacations with my family into some of the 153 national forests of our country. We were soon impressed by the natural diversity that could be seen in the national forests. Many of the features of the national parks are also to be found in the national forests, but since the national forests are publicized far less, we had to do a great deal of exploring to find these natural features. Our small project turned into a major one which has encompassed 25 years of work and pleasure and visits to each of our national forests.

I have attempted to include in this book the outstanding natural features of each forest. In so doing I have tried to indicate some of our favorite trails and campgrounds. There is something for everyone in the national forests, from the lush Quinault Rain Forest in Washington's Olympic National Forest to Alexander Springs in the heart of the subtropical Ocala National Forest in central Florida, from the brilliant red rock formations of the Dixie National Forest in Utah to the limestone rocks of New Hampshire's White Mountain National Forest, from the placid lakes wafted by cool breezes in the Superior National Forest of Minnesota to the dry, sandy deserts in Arizona's Tonto National Forest.

The national forests are home to endangered and threatened plants and animals, to all kinds of prior volcanic phenomena, and to numerous remnants of earlier cultures. Unforgettable in the national forests are the natural rhododendron gardens on Roan Mountain in the Pisgah National Forest, the great sand dunes along the Pacific coast in the Siuslaw National Forest, the imposing Chinese Wall in the Lewis and Clark National Forest, the alpine gardens on Mt. Washington in the White Mountain National Forest, the oldest living tree in the world in the Inyo National Forest, as well as giant sequoias, insect-eating plants, saguaros, palms, grizzlies, alligators, coppery-tailed trogons, and even a handful of American caribou.

Most of our national forests, unlike national parks, are sprawling land masses that often surround several parcels of private property. Imposing lodges, cabins, and eating facilities are usually not a function of national forests, but there are always plenty of nearby towns and communities to supply your needs for food, lodging, and other necessities.

National forests are a part of our public heritage—places where you can camp, picnic, hike, sightsee, hunt, fish, boat, canoe, raft, swim, ride horseback, or partake in winter sports activities. This book is intended to guide you in those endeavors.

Although the national forests are scattered throughout the United States, there are some states where there are no national forests. Among these are Massachusetts, Connecticut, Rhode Island, New Jersey, Delaware, Iowa, North Dakota, Kansas, and Hawaii. A small unit of the Green Mountain National Forest, located in New York State, and the Caribbean National Forest in Puerto Rico, both administered by the U.S. Forest Service, are experimental and are therefore excluded from this guide.

I wish to thank all those forest service personnel who provided me with maps and other information about our forests. I am particularly grateful to Mr. Charles Hendricks, supervisor of the Caribou National Forest, and Dr. William Hopkins of the Deschutes National Forest for the extra special things they did for me during the study. I also appreciate the input from a number of my students who have walked some of the trails and camped in some of the areas I was not able to get to. I extend my deepest thanks to Lea Guyer Gordon, my editor at Congdon & Weed, for the many hours she spent with the manuscript. Finally, I express my warmest appreciation to my family, Beverly, Mark, Wendy, and Trent, who uncomplainingly followed me up and down many forest trails. To my wife, Beverly, who typed the manuscript several times, I am truly grateful.

Robert H. Mohlenbrock
September 1983

CONTENTS

NORTH CENTRAL

SOUTH CENTRAL

NORTH ROCKIES

CENTRAL ROCKIES

SOUTHWEST

PACIFIC NORTHWEST

PACIFIC SOUTHWEST

EAST

There are only six national forests in the entire eastern region of the United States; these are the focus of this chapter. Those forests discussed are the White Mountain, which lies mainly in New Hampshire but extends into Maine; Green Mountain in Vermont; Allegheny in Pennsylvania; Monongahela in West Virginia; Jefferson in Virginia; and George Washington in Virginia and West Virginia.

White Mountain National Forest

Location: 686,763 acres in east-central New Hampshire and west-central Maine, near Berlin, Gorham, North Conway, and Littleton. U.S. Highways 2, 3, and 302 serve the area. Maps and brochures available at the district ranger stations in Bethlehem, Gorham, Plymouth, and Conway,

New Hampshire, and Bethel, Maine, at the Passaconaway Information Center west of Conway, or by writing the Forest Supervisor, 719 Main St., Laconia, NH 03246.

Facilities and Services: Campgrounds (some with camper disposal facilities); picnic grounds; visitor's center with exhibits and information at Passaconaway; backpacking shelters.

Special Attractions: Mt. Washington, highest mountain east of Rockies and north of Great Smoky Mountains; Alpine habitats above timberline with unusual plant life; stands of virgin red spruce; unmatched autumn foliage; two wilderness areas.

Activities: Camping, picnicking, hiking, backpacking, hunting, fishing, scenic driving, off-road vehicle activities, snowshoeing, cross-country skiing, snowmobiling.

Wildlife: Endangered American peregrine falcon; moose, and black bear.

ANY TRIP to the White Mountain National Forest must center around Mt. Washington, one of the highest mountains in the United States, and the other lofty peaks of the Presidential Range. You can see these mountains from many vantage points, but hiking into the area is the best way to experience them. This author's favorite jumping-off place is at Pinkham Notch along State Highway 16. Several trails lead from here into the 5,600-acre Pinkham Notch Scenic Area. For a short hike, we like the trail to sparkling Crystal Cascade, one of the more exciting waterfalls in the forest. If you are going to try to conquer Mt. Washington, it is a good idea to take a side trip along the way to the Alpine Garden where a remarkable assemblage of arctic flowering plants are found—all plants that generally do not grow elsewhere in the northeastern United States. Because of the extremely short growing season, these plants have to compress their entire life cycle into a two-month period. Rarest of the alpine plants is the dwarf yellow cinquefoil, one of our nation's endangered species. Other extremely rare plants, most of which are under review by the United States Fish and Wildlife Service for possible listing as endangered species, include Peck's yellow-flowered avens, the blackberrylike cloudberry, two kinds of eyebrights that are related to paintbrushes, and the alpine cudweed. Botanists have catalogued 110 different kinds of plants in the Alpine Garden, 75 of which are true alpine plants that live only above the timberline.

South and west of the summit of Mt. Washington on the Crawford Path is the famous Lake-of-the-Woods. This one-acre lake, at 5,050 feet above sea level, is the highest alpine lake east of the Rocky Mountains. You will want to pause for a moment and contemplate this stark and serene place.

In the spring of the year, skiers may hike two and a half miles from Pinkham Notch to Tuckerman Ravine, a huge glacier-carved amphitheater that serves as a basin for wind-blown snow from the Presidential peaks. The only time to ski the ravine is from late March to early June, but be advised that there are no ski lifts in the area.

Although some of the trails in the Pinkham Notch Scenic Area lead you away from the hustle and bustle of human activities, they do not compare with the isolation you experience while hiking in the Great Gulf Wilderness just north of Pinkham Notch and across the auto road to Mt. Washington. This 5,552-acre wilderness is a large glacial valley located between Mt. Washington and the northern peaks of the Presidential Range. The series of trails through rugged terrain climb from 1,700 feet above sea level to the summits of Mt. Madison, Mt. Adams, Mt. Clay, and Mt. Washington, all above 5,300 feet. Near one of the summits is a sparkling body of water known as Star Lake. Several areas above the timberline again give you a chance to observe diminutive alpine plants.

The second wilderness area in the White Mountain Forest is the 20,000-acre Presidential Range–Dry River Wilderness. Most of the trails are used as southern routes to the summit of Mt. Washington.

Another area of the forest you will want to explore is the rugged 18,000-acre Lincoln Woods Scenic Area, best reached by a trail from Wiley House along U.S. 302. This is one of the best areas in the forest to study plant and animal diversity. Between Ethan Pond and Shoal Pond is a high elevation bog that harbors rare plants such as Pursh's goldenrod, a gorgeous turtlehead, and even one kind of pitcher plant. In the same boggy area are several birds usually encountered in far northern regions, such as the boreal chickadee, pine grosbeak, Lincoln sparrow, spruce grouse, and crossbill.

There are several other areas in the forest that are worth a visit. One is the Gibbs Brook Scenic Area, lying east of the Ammonoosuc River and U.S. 302 and north of Crawford Notch State Park, where an extensive stand of virgin red spruce occurs. Associated with the spruce to form a gorgeous woods are balsam fir and yellow and paper birches. As you continue to climb east toward the summit of Mt. Pierce, the red spruce becomes stunted, and a subalpine community of dwarf vegetation flourishes.

Another area worth seeing is the Snyder Brook Scenic Area where three sets of waterfalls sparkle and cascade amid a setting of red spruce, hemlock, birches, and maples. Follow an easy mile-long trail south from Ravine House on U.S. Route 2 to the scenic area.

Perhaps this author's favorite out-of-the-way area is the Sawyer Ponds Scenic Area. Sawyer Pond, a 46-acre, 100-foot deep body of crystal clear water, is surrounded on all sides by towering stone cliffs that include the spectacular round-topped Owls Cliff. Red spruce and hemlock contrast beautifully in growth form and color with several species of hardwoods. Graceful white birches add a delicate touch. Sawyer Pond is reached via a one-mile trail from the Sawyer River Road west of Highway 302.

Of more difficult access are the Lafayette Brook Scenic Area and the Greeley Ponds Scenic Area. Lafayette Brook is located on the arduous-to-climb west slope of Mt. Lafayette. From one end of the scenic area, Mt. Lafayette's bare and formidable granite summit overshadows a forest dominated by red spruce and fir. Upper and Lower Greeley Ponds are small, shallow bodies of deep blue water surrounded by densely forested green cliffs, and mountain peaks. A hike of one and a half miles from the Kancamagus Highway brings you to this arresting area.

If you are truly adventurous, seek out Carter Ledge on Mt. Chocorua for a glimpse of a small stand of the uncommon jack pine. The going is tough at times, but the end result is worth the effort.

You can also enjoy much of the White Mountain National Forest without exerting yourself. Kancamagus Highway, for example, has several scenic overlooks and bisects the forest from east to west. Along the highway is the geologically interesting Rocky Gorge Scenic Area, where the Swift River falls 20 feet into a rock-bottomed gorge worn with numerous potholes. You can enjoy this gorge from the comfort of your car, or you can walk a few feet from your car and cast a line into the river and hope for brook trout. The Sabbaday Falls are an easy walk of less than one mile from the Kancamagus Highway. Midway along the highway is the Passaconaway Information Center, where forest service personnel hand out literature and answer questions. The center is housed in the historic George House, a frame building built in 1810 by one of the first settlers in the region.

Another easily accessible feature is the Glen Ellis Falls, where spectacular rushing waters drop 64 feet along the east side of Route 16 in Pinkham Notch.

If you still have time, drive into the small part of the forest that extends into the state of Maine and take the Patte Brook Auto Tour. Pick up a leaflet at the Bethel ranger station before starting. The Patte Brook area is designed to demonstrate some of the activities that go on in a national forest. A favorite stop is number 9, where a short walk takes you to a point overlooking the upper end of Patte Brook Marsh. If you are patient, you should see waterfowl and other wetland animals.

You can enjoy the White Mountain Forest any time of year, but the autumn coloration is truly something you never forget, as in a number of other places in New England.

Green Mountain National Forest

Location: 293,384 acres in south-central Vermont. U.S. Highway 7 and State Routes 73 and 145 serve the forest. Maps and brochures available at the district ranger stations in Manchester Center, Middlebury, and Rochester, or by writing the Forest Supervisor, 151 West St., Rutland, VT 05701.

Facilities and Services: Campgrounds (some with camper disposal service); picnic areas with grills; boat launching area.

Special Attractions: Breathtaking Texas Falls; Lye Brook Wilderness; Bristol Cliffs Wilderness.

Activities: Camping, hiking, picnicking, hunting, fishing, swimming, boating, skiing, snowmobiling.

Wildlife: Endangered or threatened bald eagle and American peregrine falcon.

THE GREEN MOUNTAIN NATIONAL FOREST runs like a backbone down the center of Vermont from near Warren to the Massachusetts state line. The mountains, worn down to rounded ridges because of their antiquity, bear a dense mantle of vegetation. The deep green color of the forest during the summer accounts for the common name of the region. Punctuating the forest are nostalgic Vermont villages, picturesque white-steepled churches, and an occasional covered bridge.

Texas Falls north of Rochester has to be on the itinerary of a forest visit. This glacier-carved gorge is climaxed by Texas Falls themselves, which dramatically drop in a short series of low falls and cascades. A trail through the hardwood-conifer forest crosses a bridge above the falls and affords a panoramic view of the cascades.

If this kind of tranquility is not enough for you, hike to the Lye Brook Wilderness where you can meditate in solitude. You can enter this wilderness from several access points. A part of the scenic Long Trail passes along the eastern edge of Lye Brook.

Another secluded region is the Bristol Cliffs Wilderness in the extreme northwest corner of the forest. This is a scenic area of exposed cliffs and rocky streams. The setting of Gilmore Pond in the south end of the wilderness is particularly pleasant.

If you are a hiker, you will enjoy Green Mountain as hiking brings you close to the forest. Long Trail, which runs the entire length of Vermont, has 80 miles in the Green Mountain National Forest. The trail has a number of excellent shelters along the way.

We were also fascinated by the words "White Rocks" on the forest map, and decided to investigate what this meant; we were not disappointed. South and east of Wallingford, and accessible via the Long Trail, lies the White Rocks, a scenic outcrop of sparkling white quartzite rock. After a short climb to the white rock area, we paused to look at the surrounding hills as they spread in every direction. There are boulder caves known as the Ice Beds at the bottom of the cliffs, where ice persists most of the summer. The White Rocks cliffs are a good place to spot the endangered American peregrine falcon. Another nice trail, slightly longer and a little more strenuous, leaves Highway 11 and climbs to the summit of Bromley Mountain. This mountain is also a famous winter recreation center for skiing. More than half the visitors to the Green Mountain National Forest come during the winter when several ski areas and miles of snowmobile trails await them. Cross-country skiing and snowshoeing are also popular.

Another fascinating attraction is Mount Horrid, a spectacular group of cliffs conspicuous from State Route 73. The area is special in that there are more than a dozen very rare kinds of plants on the mountain, including the delicate white mountain saxifrage and the white-flowered, large-leaf avens of the rose family.

We further enjoyed the Cape, a 118-acre stand of mature hardwood trees. Some of the large trees include a yellow birch with a trunk diameter of 38 inches, a red spruce whose trunk diameter is 24 inches, and a sugar maple with an 18-inch trunk diameter. Forest service personnel estimate the sugar maple to be more than 125 years old.

The poet Robert Frost has been commemorated by the Robert Frost Memorial Trail which traverses for eight-tenths of a mile in a secluded woods along the South Branch of the Middlebury River. The trail is located near Ripton.

The best developed campground is at Hapgood Pond. Located in a forested setting, Hapgood Pond is suitable for boating. A swimming beach is provided as well.

Allegheny National Forest

Location: 500,000 acres in northwestern Pennsylvania, extending from the New York state line near Warren, Pennsylvania, south for 45 miles to Ridgway. U.S. Highways 219 and 62 follow the eastern and western boundaries of the forest, and U.S. 6 and Pennsylvania 66 are the major interior highways. Maps and brochures available at the district ranger stations in Bradford, Marienville, Ridgway, and Sheffield, at the Kinzua Point Information Center near Warren, or by writing the Forest Supervisor, Spiridon Building, Warren, PA 16365.

Facilities and Services: Campgrounds (some with camper disposal facilities and some accessible only by boat); picnic grounds with grills; boat launching facilities; visitor's center with exhibits at Kinzua Point Information Center.

Special Attractions: Two rare tracts of virgin forest (Tionesta Forest and Heart's Content Forest); rugged precipitous cliffs with spectacular overlooks along the Allegheny River; rushing trout streams; gorgeous autumnal foliage.

Activities: Camping, picnicking, hiking, hunting, fishing, boating, canoeing, sailboating, swimming, waterskiing, nature study, snowmobiling, cross-country skiing, snowshoeing, ice fishing.

Wildlife: Endangered or threatened bald eagle and American peregrine falcon; black bear, bobcat.

IMAGINE YOURSELF standing in the midst of a virgin forest on the Allegheny Plateau in upper Pennsylvania. Towering hemlocks and giant smooth-barked beech trees, whose tops touch in a way that blocks the sun's rays from penetrating the silent forest floor, rise to heights of more than 100 feet. The trees in this forest have been unscathed by the woodman's axe. You can experience this solemn scene in the Tionesta Scenic Area, a 2,000-acre tract located in the heart of the Allegheny National Forest about seven miles west of the village of Kane. Beneath the cathedral forest of hemlocks are perfect specimens of black and yellow birches, sugar and red maples, white ash, black cherry, basswood, and tulip poplar, forming a habitat rarely seen today in the northeastern United States. Even the uncommon cucumber magnolia, whose interesting fruits resemble small green cucumbers, is present in the Tionesta. Also growing on the forest floor under haunting shadows cast by the dense shade are an abundance of ferns and wild-flowers such as the pink lady's slipper, the small, purple-fringed orchis, the white doll's-eyes, the delicate starflower of the primrose family, and the mountain sorrel. There are, in addition, many delightful song-birds, including the black and white warbler, Parula warbler, phoebe, chickadee, wood pewee, flycatcher, woodpecker, and the wood thrush, whose melodious calls interrupt the stillness of the forest.

Visitors can also see a small stand of virgin white pine

and hemlock, intermingled with beech and black cherry, at Heart's Content, a nearby scenic area traversed by two half-mile trails marked with botanical signposts along the way.

The biggest attraction for many visitors to the Allegheny is the recreation area made possible when the Kinzua Dam was built on the Allegheny River, impounding the water to the north into the scenic Allegheny Reservoir. Boating and waterskiing are popular, and fishermen can look for a catch of smallmouth bass, walleye, northern pike, muskellunge, channel catfish, crappie, or yellow perch. Smaller tributaries that feed the reservoir are lucrative grounds for trout.

Following spring rains, you might want to take a small boat from near the Tionesta Campground up the scenic Tionesta Creek which meanders through a dense canyon.

Long-distance hikers can choose from several challenging trails, including nearly 100 miles of the North Country Trail which zigzags through most of the Allegheny. There is a good picnic spot at Jakes Rock. Perched on a cliff nearly 600 feet above the Allegheny River, Jakes Rock is in an area of rugged beauty. Several trails parallel the cliff's rim, so be careful not to stray too near the edge.

As for camping, you can select a variety of possibilities. Handsome Lake, on the east edge of the Allegheny Reservoir, can be reached only by boat. One of our favorite secluded spots is the small campground at Kelly Pines in the heart of the forest, where sweet-scented pines waft gracefully in warm summer breezes.

If you can get to the Allegheny in October, be prepared for a vivid show of autumnal coloration provided by the changing leaves of the sugar maple, red maple, birch, tulip poplar, and white ash.

Monongahela National Forest

Location: 846,000 acres in eastern West Virginia, between White Sulphur Springs and Elkins. U.S. Highways 33, 219, and 250 cross the forest. Maps and brochures available at the district ranger stations in Parsons, Rockwood, Bartow, Marlinton, Petersburg, and White Sulphur Springs, at the Spruce Knob-Seneca Rocks National Recreation Area Visitor's Center, at the Cranberry Mountain Visitor's Center, or by writing the Forest Supervisor, USDA Building, Sycamore St., Elkins, WV 26241.

Facilities and Services: Campgrounds (some with camper disposal service); picnic areas with grills; boat launch areas; beaches; visitor's centers.

Special Attractions: Unusual group of northern plants and animals in the Cranberry Glades; Seneca Rocks; Otter Creek Wilderness; Sinks of Gandy, a disappearing creek.

Activities: Camping, picnicking, hiking, backpacking, hunting, fishing, swimming, boating, canoeing, hawk watching, nature study.

Wildlife: Black bear.

OF THE MANY FEATURES in the Monongahela National Forest, the famous Cranberry Glades are our favorite. Long before we ever had the opportunity to see them, we read about the bogs and the unusual northern plants and animals that lived there. We suggest stopping at the Cranberry Mountain Visitor's Center where you will receive a good orientation on the area. Then, before heading for the Cranberry Glades, we suggest walking a short nature trail near the center to get an occasional glimpse of the backcountry.

You will find that the Cranberry Glades are located at an elevation of about 3,400 feet. There are 720 acres of glades: 400 acres are a dense thicket of speckled alder, 200 acres are bog forest, while the remaining 120 acres are open glade. In reality there are five separate glades, each covered with sphagnum or peat moss. In some places the peat may be as much as 11 feet deep. Among the rare northern plants found are buckbean, an herb related to gentians; the yellow clintonia and green hellebore of the lily family; marsh marigold, an aquatic buttercup; the delicate, six-inch-tall dwarf dogwood; the ill-scented muskflower of the snapdragon family; and bog rosemary. Each of these plants is located hundreds of miles south of where they ordinarily grow. There are, of course, cranberries, both large and small, as well as insect-catching sundews and pitcher plants. Northern birds that find their southernmost breeding area in the Cranberry Glades include the hermit thrush, olive-backed thrush, purple finch, northern water finch, Nashville warbler, mourning warbler, swamp sparrow, and alder flycatcher. A boardwalk across a part of the glades allows visitors to observe this unusual biological community without disturbing the fragile habitat and without getting their feet wet.

North of Cranberry Glades is what is called Cranberry Backcountry, a 53,000-acre tract of near-wilderness ranging from Williams River at 2,350 feet above sea level to the 4,625-foot summit of Black Mountain. Red spruce covers most of the higher land, while at lower elevations one discovers black walnut, tulip poplar, white oak, red oak, basswood, white ash, hickory, and maple. Williams River, by the way, offers superb white-water canoeing between Tea Creek and Three Forks.

While in cranberry country, we suggest you drop down to the Falls of Hills Creek Scenic Area to see the impressive three rushing falls of Hills Creek. You will hike down a 250-foot narrow ravine. The upper falls drop 25 feet, the middle falls drop 45 feet, and at the lower falls the drop is 63 feet. Observation platforms provide excellent places to view the falls, but beware that the wooden stairways along the trail can become very slippery when wet.

Another area we recommend exploring is the high plateau called Huckleberry Plains where huckleberries and blueberries dominate the vegetation. Red spruces are also present, but the harsh climate keeps them stunted. Within the plains the best known area is 10,000 acres that have been designated the Dolly Sods Wilderness. Forest Road 75 skirts the eastern edge on its way to Bear Rocks, making the area readily accessible. Bear Rocks, at the edge of the Allegheny Front Mountain, are, in fact, masses of bare white rocks that are fun to scramble over. Bear Rocks is also an excellent place to watch hawk migrations in the autumn. At Dolly Sods you can also pick the huckleberries when they are ripe. Because the area contains numerous swampy depressions, you must always watch where you step. These depressions are home to cranberries, mountain holly, and speckled alder.

To appreciate the size of the mountains in the Monongahela, drive to the top of 4,860-foot Spruce Knob, the highest spot in West Virginia. On a clear day, from the 22-foot observation tower, you can see North Fork Mountain to the east and Middle Mountain to the west across Seneca Valley. Spruce Knob itself is bleak and cold, with harsh winds always blowing. One notes that red spruces are only 10 to 20 feet tall and have branches only on one side, even though they may be 75 years old. The trail on the summit of Spruce Knob passes eye-catching vegetation, including the pink lady's-slipper orchid. You can picnic on the mountaintop if weather permits. Nearby Spruce Knob Lake is stocked with trout and has a campground.

The most familiar landmark in the Monongahela is Seneca Rocks, a monolith of white sandstone that rises 1,000 feet above North Fork. Hundreds of persons climb to the summit of Seneca Rocks each year. Most go up the back side on a very steep path; others bring their mountain climbing apparatus and scale the front. Either way you get there, the summit provides an unmatched view of the Spruce and Allegheny Front mountains. An outstanding white sandstone formation known as Champe Rocks can be seen near Seneca Rocks. Another spectacular rock formation that juts 300 feet into the air is Eagle Rock near the South Branch of the Potomac.

Smoke Hole is the picturesque name given to the valley of the South Branch of the Potomac where the river enters a narrow canyon and flows for 20 miles through rough terrain. The Smoke Hole road follows this canyon for some distance before terminating at a secluded campground.

Spruce Knob, Seneca Rocks, Champe Rocks, Eagle Rock, and Smoke Hole are all included in the Spruce Knob–Seneca Rocks National Recreation Area. A helpful visitor's center providing information and exhibits on the area is located at the junction of U.S. 33 and West Virginia 28 at Mouth of Seneca.

We consider the best wilderness experience in the Monongahela the Otter Creek Wilderness. Otter Creek, a beautiful, fast-flowing stream, cascading over numerous rock shelves, flows through this densely forested area. One feels in the depths of this wilderness a complete sense of isolation from the 20th century. Ferns are plentiful, as are rhododendrons and mountain laurel. An interesting area along Otter Creek is Moore Run Bog, a spongy, high altitude bog with a number of unusual northern plants, such as the insectivorous sundews and delicate orchids. If you take the trail from Mylius Gap north to the summit of Shavers Mountain, you will pass a small stand of untouched hemlock.

For another inspiring moment, try the short trail at Gaudineer Scenic Area that winds among 130 acres of very mature red spruce, many over 100 feet tall. At an elevation of 4,445 feet, this is good bird-watching country, with several species of warblers usually present.

There are two further remote areas we suggest. One is Blister Swamp, a 40-acre stand of balsam fir (known locally as blister pine) with a small cold-water spring, where conditions are ideal for several northern species of plants, including twinflower, white monkshood, golden saxifrage, the alpine enchanter's nightshade, goldthread, and New York fern. The other remote area is Sinks of Gandy, named for Gandy Creek, a stream of water that completely disappears to the eye beneath Yokum Knob only to emerge again 3,000 feet away.

If you are looking for a comfortable campground, try the one at the Stuart Recreation Area. The campsite is in a picturesque setting of majestic pines, and nearby Shavers Fork River is good for fishing and swimming. The campground, which also has a nature trail nearby, is the starting point for the Stuart Memorial Drive, a primitive mountain road that leads to Bickel Knob, Bear Haven, and eventually Alpena Gap.

Jefferson National Forest

Location: 670,000 acres in west-central Virginia, from the James River southwest nearly to the Tennessee state line. Interstates 91 and 77 and U.S. Highways 460, 21, and 52 are the major highways. Maps and brochures available at the district ranger stations in Blacksburg, Wise, Natural Bridge Station, Marion, New Castle, and Wytheville, at the Mt. Rogers National Recreation Area headquarters, or by writing the Forest Supervisor, 210 Franklin Road SW, Roanoke, VA 24001.

Facilities and Services: Campgrounds (some with camper disposal service); picnic areas with grills; boat launching areas; information center at Mt. Rogers National Recreation Area.

Special Attractions: Mt. Rogers National Recreation Area; only location in the world for Peters Mountain mallow; James River Face Wilderness; iron furnaces.

Activities: Camping, picnicking, horseback riding, backpacking, boating, hunting, fishing, swimming, nature study.

Wildlife: Threatened bald eagle; black bear and bobcat.

MOST VISITORS to the Jefferson National Forest sooner or later make their way to the Mt. Rogers National Recreation Area, and little wonder, for this 100,000-acre region of pleasant climate and diverse vegetation is the focal point of the forest. It is a hiker's paradise. There is a range of choices from a short nature trail at Grindstone Campground to the 38½-mile hike from Whitetop Mountain to Iron Mountain via Whitetop-Laurel Creek. Our favorite medium-range hike is the 11½-mile Whitetop-Laurel Creek Circuit Trail. One begins at the Beartree Campground, then follows the Beartree Gap Trail for about ½-mile until it crosses the Appalachian Trail. There one turns right and follows the Appalachian Trail for 4 miles. The next turn is onto an old abandoned bed of the N and W Railroad through Taylors Valley and along Whitetop-Laurel Creek. The trail crosses the creek several times on trestles, the most exciting being the 500-foot-high trestle at Creek Junction.

There is a vast variety of nature in the Mt. Rogers National Recreation Area as well as hiking. In areas above 3,500 feet, there are lovely open meadows dotted with spruce and fir, beech, birch, and maple, and a thick growth of herbs. At lower elevations are dense forests of tulip poplar, white oak, and red oak. You may want to try and locate a specimen or two of the Virginia round-leaf birch, the only tree on the United States list of endangered species. It looks very much like any other birch, except for its round leaves. Keep in mind that it is illegal to collect from these trees. Birdwatchers might be rewarded with a glimpse of the sharpshinned hawk, the American kestrel, or Bewick's wren. You may also catch sight of the northern flying squirrel.

A myriad of recreational activities is available at Mt. Rogers. At the Beartree Recreation Area, you can swim in a 14-acre lake, fish for trout in nearby Straight Branch, or enjoy an occasional interpretive program in the amphitheater. An extensive horse trail runs most of the length of the recreation area, beginning near Ivanhoe, Virginia, and extending for nearly 50 miles to the west of Mt. Rogers. The trail climbs to Chestnut Knob and follows the ridge until it drops down to cross Francis Mill Creek. It continues over several ridges and valleys and follows numerous picturesque streams. Camping areas with horse facilities are found along the trail. If you do not have your own horse, you may rent one from the Fiarwood Livery on Virginia Route 603.

Hurricane Campground is one of the more secluded camp-grounds, hidden near two trout streams among the Iron Mountains. If you only wish to picnic, try the Shepherds Corner area located on a scenic bluff high above the New River.

There is much more in the remaining 570,000 acres of the Jefferson National Forest besides the Mt. Rogers National Recreation Area. Among the choicest regions is the Mountain Lake Wilderness Study Area northeast of Pembroke. These nearly 12,000 acres of wild and scenic country encompass a variety of natural habitats. There are several spruce bogs where stands of the once common red spruce can still be seen, along with rhododendrons and a variety of unusual herbs. Towering hemlock groves are present, with some of the trees more than 300 years old. One stand is in the upper reaches of War Spur Branch. Throughout the Mountain Lake area are assemblages of stately deciduous trees. The upper slopes of Johns Creek Valley have a splen-did forest of white oak, red oak, black oak, chestnut oak, basswood, American elm, tulip poplar, sugar maple, beech, and sourwood. In the autumn the sugar maples and sour-woods are downright showy in their brilliant orange and crimson coloration. On top of Johns Creek Mountain and Salt Pond Mountain is a dense entanglement of bear oak as well as shrubby members of the heath family, giving this habitat the appearance of a western chaparral, a community of plants with a brushy appearance. Of the several imposing cliffs in the Mountain Lake area, Wind Rock and White Rocks are the most spectacular. This roadless area, with rugged trails and primitive campsites, is a good place to seek solitude. Only an occasional shy black bear may in-terrupt your peace.

Peters Mountain is another wild region in the Jefferson National Forest. Outcrops of broken sandstone make the area particularly rugged. If you get to Peters Mountain, look for the rare, large-flowered Peters Mountain mallow. This is the only location in the world where this plant grows. This hollyhock-like plant is being considered by the United States Fish and Wildlife Service for its endangered species list. Still another wild region is the James River Face Wil-derness between Lexington and Bedford, a rugged 8,700-

acre roadless area in the northeast corner of the forest. It is only for hardy souls.

In addition there are several native trout streams in the forest. Roaring Fork Creek is one of the more turbulent streams, with boulders covered with dense vegetation and several scenic waterfalls. The most spectacular waterfall is the Cascades, which drops 60 feet into a horseshoe-shaped pool. To get to this falls, take the four-mile, round-trip Cascades National Recreation Trail.

During the middle of the 19th century, the Jefferson National Forest was the center of the iron industry, and some bits of evidence still remain. For example, Roaring Run Furnace, abandoned since 1865, still stands about one mile from Craig Creek. Remains of the Glenwood Furnace can also be seen. To get a feel for the old iron industry area, take the Fenwick Forest nature walk, which provides a pleasant diversion for a stroll into history.

George Washington National Forest

Location: One million acres in northwestern Virginia and adjacent West Virginia. Major access routes are Interstates 64 and 81 and U.S. Routes 220, 250, 33, and 211. Maps and brochures available at the district ranger stations in Staunton, Bridgewater, Covington, Edinburg, Buena Vista, and Hot Springs, at the Massanutten Visitor's Center near New Market Gap, or by writing the Forest Supervisor, 210 Federal Building, Harrisonburg, VA 22801.

Facilities and Services: Campgrounds with camper disposal services; picnic areas with grills; canoe camp facilities; swimming beaches; boat launch areas.

Special Attractions: Iron furnaces; old growth stand of hemlocks; boggy area known as a muskeg; historic mansion.

Activities: Camping, picnicking, hiking, hunting, fishing, canoeing, boating, swimming, nature study.

Wildlife: Black bear.

SHENANDOAH AND MASSANUTTEN are names that recall the turbulent battles of America's Civil War more than a century ago. They are also the names of mountains in the George Washington National Forest, a 1-million-acre expanse of forested and mountainous public land that parallels the Virginia–West Virginia state line from near Winchester in north-central Virginia almost to White Sulphur Springs in southeastern West Virginia. Today these mountains offer an unmatched serenity and the lovely beauty of a soft green mantle of vegetation; yet you can see the remnants of heavy and often violent past military activity. One of the forest's attractions is its iron furnaces, where ore was once smelted for cannons, guns, and other needs of war. Some are now restored although there are others that are reduced to rubble. Even if you are not a Civil War buff, you might enjoy a visit to the old Confederate breastworks which are north of U.S. Highway 250 near Ramseys Draft. Between Strasburg and Front Royal is the site of Powells Fort.

Before exploring the forest, you will want to know that the George Washington is divided into three separate units. You can get a good picture of things by starting in the easternmost segment, between Front Royal and Harrison-burg. This narrow tract, dominated by Massanutten Mountain, is flanked near either end by restored iron furnaces. The Elizabeth Iron Furnace is at the north end of Fort Valley. Pig Iron Trail, which meanders through the area, has numbered trail markers that correspond to interpretive remarks in a guide leaflet. All that is left of the furnace, which was abandoned in 1888, is part of the stack. A short distance beyond, you can see the location of the abandoned town of Elizabeth. The Elizabeth Furnace campground is nearby. The Catherine Furnace, another relic of the 19th century, recently has been restored. It is a few miles south of New Market Gap. New Market Gap, where U.S. Highway 211 crosses Massanutten Mountain, is a particularly pleasant place to picnic, and we heartily recommend it. Nearby is the Massanutten Visitor's Center.

You should spend time exploring the Massanutten. The region is outstanding botanically, because of the unusual assemblage of plants. Starflower, a white-flowered member of the primrose family, usually grows in woods far to the north, but it also occurs here along a small stream at the base of a steep-walled gorge. Growing with it is the northern yellow violet, another plant far south of its usual range. While you are in this area, look for the bright yellow marsh marigold, the creeping, red-fruited partridge berry, and the sweet-scented, delicate, showy orchis.

If you do not mind getting your feet wet, seek out the Massanutten muskeg, a special attraction in the forest and an unusual habitat in Virginia along Peter's Mill Run in the vicinity of the Little Fort Campground. Muskeg, a word meaning "trembling earth" to American Indians, is a quaking bog. Since this 2,000-square-foot muskeg is fragile, observe the unusual plants in it from the adjacent forest. Plants you will likely see are the marsh fern, cotton sedge, rose pogonia orchid, little twayblade orchid, and the insectivorous round-leaved sundew.

Before leaving this section of the George Washington National Forest, take a moment to stop by the Camp Roosevelt Recreation Area, which has been developed around the site of the first CCC (Civilian Conservation Corps) camp built in the United States in the 1930s. You can camp and picnic in the area. Nearby, there are two canoe camps along the Shenandoah River that you can use as your base for canoeing.

The best way to see the southeastern unit of the George Washington is to drive the Blue Ridge Parkway or hike the Appalachian Trail, both of which run the entire length of this section. The spacious campground at postcard-pretty Sherando Lake, with its adjacent swimming area, is recommended.

The largest continuous land mass of the George Washington National Forest straddles the West Virginia–Virginia line for 150 miles. Take time to hike along Ramseys Draft, a babbling, rocky stream, through a deep woods laden with wildflowers and ferns to an old-growth virgin forest of hemlock. Just a short distance southwest of Ramseys Draft is the old Confederate breastworks mentioned earlier.

A hike to the Big Schloss Geological Area on the state line a few miles north of Edinburg is also worth the effort. The Big Schloss contains several massive rock outcrops from which 360-degree views await the hiker who can make it to the top.

At Hidden Valley northwest of Warm Springs is a primitive campground crossed by Jackson River, one of the finest trout streams in the forest. The handsome and stately Warwick Mansion, built in 1850 and now a National historic Place, is here. It is well worth a stop and visit.

SOUTHEAST

The second largest concentration of national forests in the United States is in the Southeast. Their total number is 23. The forests are the Bankhead, Conecuh, Talledega, and Tuskegee in Alabama; Bienville, Delta, DeSoto, Holly Springs, Homochitto, and Tombigbee in Mississippi; Chattahoochee and Oconee in Georgia; Apalachicola, Ocala, and Osceola in Florida; Croatan, Nantahala, Pisgah, and Uwharrie in North Carolina; Francis Marion and Sumter in South Carolina; Cherokee in Tennessee; and Daniel Boone in Kentucky.

Bankhead National Forest

Location: 179,600 acres in northeastern Alabama, between Haleyville and Cullman. U.S. Highway 278 and State Route 35 cross the forest. Maps and brochures available at the

district ranger stations in Double Springs and Haleyville or by writing the Forest Supervisor, 1765 Highland Ave., Montgomery, AL 36107.

Facilities and Services: Campgrounds (some with camper disposal service); picnic areas with grills; swimming beaches; boat launching areas.

Special Attractions: Sipsey Wilderness; Pine Torch Church, oldest log church in Alabama.

Activities: Camping, picnicking, hiking, swimming, boating, canoeing, hunting, fishing, nature study.

Wildlife: Threatened red-cockaded woodpecker.

MOST OF YOUR TIME spent in the Bankhead National Forest will probably be in the vicinity of the Sipsey Branch of the Black Warrior River, because most of the spectacular canyons and gorges are in this region. In our opinion, the hike along Bee Branch from Forest Road 208 to the junction of Bee Branch with the Sipsey River is the most inspiring. The highlight of this trail is the rock-rimmed box canyon with sheer vertical walls rising to 100 feet or more. The canyon contains a never-before-lumbered hardwood forest where evergreen hemlocks and mighty tulip poplars grow together. In fact, the largest tulip poplar in Alabama is in this box canyon, a tree with a trunk diameter of more than 6½ feet. You will appreciate the size of this mammoth tree if you climb down into the cove and measure yourself beside it.

Bee Branch and the area surrounding it are included in the Sipsey Wilderness. Several streams are tributary to the Sipsey River, and each of them has carved a scenic canyon and gorge. The trails along Thompson Creek, Hubbard Creek, Quillan Creek, Borden Creek, and the Sipsey River are all worth taking, for the scenery is breathtaking and the vegetation is lush.

A good place to explore the Sipsey River is from the Sipsey River Picnic Area along Cranal Road (Forest Road 26). A well-beaten nature trail skirts along the bases of rocky sandstone bluffs, many of them with a honeycomb surface, the result of centuries of erosion. In late April and early May, there is an astounding variety and abundance of delicate wildflowers and fragile ferns. The Sipsey River can also be canoed.

Other areas we like in the Bankhead are the Natural Bridge, where a short but rather strenuous trail winds in an area of picturesque rocky bluffs to a natural sandstone bridge, and Brushy Lake. You can camp or picnic along the shore of this serene lake, swim and fish in it, and take an easy trail that passes beneath neat rock bluffs. If you prefer a less secluded campground, use the one called Corinth on the edge of the 21,000-acre Lake Lewis Smith.

Before leaving the Bankhead National Forest, drive north out of Grayson on Forest Road 245 and then 246 until reaching the ancient Pine Torch Church, one of the forest's special attractions. The one-room log building was constructed in 1808 and is thought to be the oldest log church in Alabama. Burning pine knots were placed in the rafters to provide light—which accounts for the name of the church.

Conecuh National Forest

Location: 83,000 acres in southern Alabama between Andalusia and the Florida state line. U.S. Highway 29 and State Route 55 are the major access roads. Maps and brochures available at the district ranger station in Andalusia or by writing the Forest Supervisor, 1765 Highland Ave., Montgomery, AL 36107.

Facilities and Services: Campground with modern facilities at Open Pond; picnic areas with grills; swimming areas.

Special Attractions: Open Pond; Blue Pond; hiking trails.

Activities: Camping, picnicking, hiking, swimming, hunting, fishing.

Wildlife: Endangered or threatened red-cockaded woodpecker, bald eagle, American alligator, eastern indigo snake.

CROWDED AGAINST the Florida state line, the Conecuh National Forest in southern Alabama is in the flat-to-gently-rolling sandy coastal plain. Low, wet areas are interspersed here and there. Lazy streams, such as Panther Creek and Blackwater Creek, flow through areas of dense vegetation. Many cypress-bordered ponds are in the forest. Much of the Conecuh is dominated by longleaf pine, although several hardwood swamps are present.

Most of the activities in the Conecuh center around two natural lakes. One is Open Pond, a 50-acre lake that has a developed campground, picnic area, and swimming beach. Another is Blue Pond, a slightly smaller lake that has a fine picnic area and a swimming beach, but camping is not permitted there. Fishing is generally good in both lakes, as well as in the much smaller Nellie Pond and Mossy Pond.

The Conecuh Trail offers hikes of varying lengths. The complete trail is about 20 miles long, but there are shorter segments that pass through interesting countryside. One and a half miles of the Conecuh Trail loop around Open Pond, and a 4-mile trail connects Open Pond with Blue Pond. There is a most interesting loop trail from Open Pond to Blue Springs, a large, picturesque natural spring. This segment of the trail is nearly 3 miles long. The longest loop trail, an 11-miler, from Blue Pond, passes by three naturally occurring small ponds—Nellie Pond, Gum Pond, and Mossy Pond. A considerable amount of wildlife can be seen along the trails, including white-tailed deer, wild turkey, raccoons, and squirrels. Keep an eye open for the red-cockaded woodpecker in old diseased pines, for American alligators in the wettest areas, and for the large eastern indigo snake, a blue-black, nonpoisonous reptile that may reach six feet long. All are on the federal endangered and threatened species list. The eastern indigo snake and the state-endangered Florida pine snake live in the burrows of the equally rare gopher tortoise. It is illegal to molest any of these animals.

Talladega National Forest

Location: 371,000 acres in two units, one east of Birmingham between Sylacauga and Piedmont, the other southeast of Tuscaloosa. Interstate 20, U.S. Highways 431, 280, and 82, and State Routes 5, 77, and 219 are the major access routes. Maps and brochures available at the district ranger stations in Centreville, Heflin, and Talladega, or by writing the Forest Supervisor, 1765 Highland Ave., Montgomery, AL 36107.

Facilities and Services: Campgrounds (some with camper disposal service); picnic areas with grills; swimming beaches.

Special Attractions: Scenic mountain drives.

Activities: Camping, picnicking, hiking, hunting, fishing, boating, canoeing, swimming.

Wildlife: Threatened red-cockaded woodpecker.

THE GREAT ATTRACTION of Talladega forest is its scenic routes. You will find that there's a bit of the Appalachians in the eastern unit of the forest between Sylacauga and Piedmont. Drive the Talladega Scenic Highway or the Skyway Motorway on either side of Cheaha State Park, and you'll see why. Although these are not the high mountains typical of the Great Smoky area, they are still big mountains, with rocky ridges, densely forested coves, and babbling brooks and streams. The Talladega Scenic Highway is a smoothly paved road, running for nearly 20 miles along the Talladega Mountain Range. There are many scenic vistas, such as the one from the rocky escarpment known as Sherman Cliff. The highway runs from Cheaha State Park to State Route 78; an extension of this road has been on the drawing board for years. If you want to take a more primitive route, drive the Skyway Motorway east from Chandler Springs. The one-lane, rocky road goes for 11 miles through rugged but picturesque mountain country. The initial ascent from Chandler Springs is steep before the road begins to follow narrow ridges.

For hikers there is one scenic trail of particular note. It is the Pinhoti Hiking Trail and it offers hikers a large serving of nature experiences. The trail wanders for 80 miles in the eastern unit of the forest. It takes one into the diverse conditions of this national forest, climbing over ridges and descending into valleys. Sometimes it passes through pine forests; other times it is beneath hardwood trees. For a while the trail veers near the edge of rock bluffs; then it enters rich, forested coves; or it may parallel rocky streams whose waters are crystal clear. To keep you from straying from the trail, the forest service has carved the likenesses of turkey tracks on some of the trees. Near the northern terminus of the trail, and a short distance south of the Coleman Lake Recreation Area, is the old Shoal Creek Church, a national historic landmark. This hand-split, hand-hewn pine log building was built between 1885 and 1890. Take time to examine the construction of this building. Coleman Lake is suitable for boating with small boats, and it has a swimming beach. Fishing is said to be good here.

A short distance southwest of the Shoal Creek Church is a lovely, secluded campground and picnic area known as Pine Glen. The scenery is superb, topped off by a sparkling mountain stream that flows through the area. Forest Highway 500, which serves the campground, is one of the prettiest in the forest.

West of Chandler Springs is Horn Mountain. If you choose to use the picnic grounds here, you will be treated to one of the most panoramic views in the state of Alabama.

The western section of the Talladega lies southwest of Tuscaloosa in an area of small mountains and rolling hills. One of the more popular places in this part of the forest is the Payne Lake Recreation Area between Centerville and Greensboro. In addition to a nice campground, pleasant picnic areas, and a swimming beach, there is a varied nature trail that climbs out of a wet bottomland forest of trees, many over 100 feet tall, to a dry ridge where the vegetation is stunted because of the harsh environmental conditions. For a while the trail follows the route of a long-ago-dismantled railroad bed that was used primarily to haul timber out of the forest. Along the trail are several kinds of pine, magnolia, flowering dogwood, tulip poplar, beech, aspen, and sweet gum. There is a chance you might see a wild turkey among the variety of wildlife here. The bobcat is also here, but you are unlikely to catch a glimpse of this elusive animal. For a pleasant picnic spot, try the one on Cahaba Mountain. From your table you will have a commanding view of the countryside.

Tuskegee National Forest

Location: 10,795 acres northeast of Tuskegee, Alabama. Interstate 65 and U.S. Highways 29 and 80 pass through the forest. Maps and brochures available at the district ranger station east of Tuskegee or by writing the Forest Supervisor, 1765 Highland Ave., Montgomery, AL 36107.

Facilities and Services: Campground; picnic area; auditorium at ranger's station.

Special Attractions: Bartram Trail; Booker T. Washington cabin replica.

Activities: Camping, picnicking, hiking, hunting, fishing.

Wildlife: Threatened bald eagle.

THE TINY TUSKEGEE NATIONAL FOREST is in the piedmont of Alabama, a region of rolling hills and sluggish streams. Lying between the mountains and the coastal plain, the Tuskegee shares characteristics of both of these provinces.

A modern ranger station facility in the heart of the forest enhances the Tuskegee National Forest and is a good place to commence a visit. You can pick up pertinent literature as well as view slide programs and films in the auditorium. You also can get a feeling of the Bartram Trail, one of the forest's major attractions, before you start your hike.

As you will discover, the Bartram Trail stretches for eight and a half miles through the pinewoods of the Tuskegee. It is a trail with few difficult segments and brings you in touch with most of the kinds of flowering plants that grow in the forest. There are several access points where you can join the trail.

At Taska, the only recreation development in the forest, you can camp at the tiny campground or picnic in a tranquil pine grove. On the site is a replica of an old log cabin similar to the one where the famous black educator Booker T. Washington was born and lived.

Hunting is permitted in the forest, and fishing is fairly good in a couple of beaver ponds and in the streams that meander through the forest.

Bienville National Forest

Location: 178,000 acres in central Mississippi around the town of Forest. Interstate 20 and State Highway 35 pass through the forest. Maps and brochures available at the district ranger stations in Forest and Raleigh or by writing the Forest Supervisor, 100 W. Capitol St., Jackson, MS 36269.

Facilities and Services: Campgrounds with camper disposal service; picnic areas with grills; swimming beaches; boat ramps.

Special Attractions: Bienville Pines; stand of virgin woods; Harrell Prairie Hill Botanical Area.

Activities: Camping, picnicking, hiking, hunting, fishing, swimming, boating.

Wildlife: Threatened red-cockaded woodpecker.

AN HOUR'S DRIVE east of Jackson, Mississippi, is a beautiful region of forested, rolling hills, many of them covered by tall pines and handsome hardwoods. Much of the region is in the Bienville National Forest.

Two areas within a mile of each other just outside the city limits of the town of Forest are worth a visit. One of these is a 167-acre virgin forest known as the Bienville Pines Scenic Area. The area contains many trees of shortleaf and loblolly pines that are more than 100 feet tall. Contrasting with these tall pines are large specimens of sweet gum and various kinds of oaks. A variety of circumstances has kept the forest free from timber harvesting. A signposted trail has been laid out through a portion of this virgin forest to add to our understanding. A couple of large pines utilized by the threatened red-cockaded woodpecker can be seen, as well as a number of trees that have been struck by lightning. Although the trail is an easy one (there are no hills here), you must be careful of the poison ivy that occurs along the trail. A small parking area is located at the beginning of the trail.

After leaving the Bienville Pines Scenic Area, drive a short distance along Forest Road 518 to the crest of a hill where a sign indicates you have reached the Harrell Prairie Hill Botanical Area. Perched on top of the hill, on either side of the forest road, is the best example of a tall grass prairie in this part of Mississippi. During the summer big bluestem and Indian grass, both tall prairie species, grow with colorful prairie wildflowers such as purple prairie clover, white prairie clover, and brilliant orange butterfly weed. There is a considerable amount of encroachment on the prairie by shrubs and weeds.

The major recreation area in the Bienville is at Marathon Lake. This scenic, 70-acre lake is surrounded by forests that give an ideal setting for camping, picnicking, and hiking. There is a swimming beach with an adjacent bathhouse, and boaters can put small boats in at the launching area.

Two very small lakes with limited camping facilities at Raworth and Shongelo are said to be excellent for bass and bream fishing. Shongelo also provides the opportunity for swimming.

Delta National Forest

Location: 59,000 acres in west-central Mississippi between Rolling Fork and Vicksburg. State Route 16 goes through the forest. Maps and brochures available at the district ranger station in Rolling Fork or by writing the Forest Supervisor, 100 W. Capitol St., Jackson, MS 36269.

Facilities and Services: Campground; picnic area with grills; boat launching area.

Special Attractions: Green tree reservoir; large bald cypress trees; bayous.

Activities: Camping, picnicking, hiking, boating, hunting, fishing.

Wildlife: Threatened red-cockaded woodpecker.

THE DELTA NATIONAL FOREST is a pleasant place to camp, but it is really a mecca for hunters and fishermen. There are large populations of white-tailed deer and squirrels, and turkeys are rather plentiful. In order to attract a variety of waterfowl, a green tree reservoir has been developed in the forest. This is a low area where water is allowed to flood for a portion of each year. Not only does the water insure an abundance of waterfowl, but it also encourages the growth of moisture-loving oaks, which respond by producing bountiful crops of acorns.

For fishermen, both the Big and Little Sunflower rivers which bisect the forest are good for bass, bream, and catfish.

The only campground is at Blue Lake, but it is a good one. The campsites are near an attractive bayou, where very large bald cypresses and other trees that can tolerate standing water are beautiful. It does not take long to get your catch for the day to the frying pan. The aroma of frying catfish, bream, or bass seems to linger in the campground and has to be one of the greatest stimulations to one's physical senses.

DeSoto National Forest

Location: 500,500 acres in southeastern Mississippi, between Laurel and Gulfport. U.S. Highways 49 and 98, and State Routes 15, 29, 63, and 67 serve the forest. Maps and brochures available at the district ranger stations in McHenry, Wiggins, and Laurel or by writing the Forest Supervisor, 100 W. Capitol St., Jackson, MS 36269.

Facilities and Services: Campgrounds (some with camper disposal service); picnic areas with grills; swimming beaches; boat launching areas.

Special Attractions: Tuxachanie Trail; unusual titi swamp; white-sand-bottomed lakes.

Activities: Camping, picnicking, hiking, boating, canoeing, swimming, wading, hunting, fishing.

Wildlife: Threatened red-cockaded woodpecker and American alligator.

IN THE SOUTHEASTERN CORNER of Mississippi is a land of cypress-lined streams, rich bottomland forests, beautiful pine woods, wide-spreading live oaks, and lakes with white sandy bottoms. Some of this region is in the DeSoto National Forest, whose two units stretch nearly to the Gulf of Mexico.

One of the best ways to become familiar with the DeSoto is to hike all or part of one of its star attractions—the Tuxachanie Trail. There are several easy accesses to this 20-mile trail, but we believe the main entrance along U.S. Highway 49 a short distance north of Gulfport is the best because of the staggering-high live oaks that arch over the parking lot. The trail winds through nearly every habitat in the forest. For a while it follows flowing streams that are encroached upon by heavy growths of vines and other vegetation. Then one comes upon a small pond which at one time served an old farm. Suddenly the forests give way to grassy openings known as savannas. An abundance of palmetto serves as a reminder that you are on the fringe of subtropical America. If you look around in these savannas, you will likely see insect-eating pitcher plants (so-called because of their shape) and narrow-leaved sundews whose sticky stems also entrap tiny insects. Occasional mounds of sand near the trail are the home of cacti. Where the trail goes over low ridges, longleaf and slash pines rise above small trees of yaupon holly and the leathery-leaved farkleberry. There are several quaint rustic bridges that keep you from wading across the streams that feed Tuxachanie Creek. Spring is a nice time to hike the Tuxachanie Trail because mountain laurel is in bloom. A time *not* to take the trail is during the hot and humid summer months.

Another special area of unusual botanical interest is the swamp of titi trees along Railroad Creek where Stone, Harrison, and Jackson counties come together. The largest southern titi tree in the world is here. This swamp is dominated by black titi and southern titi (also known as buckwheat tree and swamp cyrilla, respectively). Other interesting plants you will see are the wax myrtle, gallberry, and beauty-berry.

If you want to explore and wade in one of the many white-sand-bottomed ponds and small lakes in the DeSoto, seek out the New Augusta Pond or the four-acre lake along Miles Branch. They are adjacent to each other along U.S. Highway 98, 11 miles east of U.S. Highway 49.

For camping, we recommend the Big Biloxi Campground along the banks of the Big Biloxi River west of Highway 49 about 14 miles north of Gulfport. You can fish in the river and hike a short trail through the bottomland hardwood forest nearby.

There are 50 miles of canoeable streams along Black Creek and Beaver Dam Creek. The routes follow two of the most scenic streams in southern Mississippi. You may even see a rare American alligator basking in the sun along the shore. If you did not bring your canoe, you can rent one in the nearby community of Wiggins.

Holly Springs National Forest

Location: 147,000 acres in two units in northern and north-central Mississippi. Interstate 55 borders the western unit; U.S. Highway 78 diagonally crosses the eastern unit. Maps and brochures available at the district ranger station in Holly Springs or by writing the Forest Supervisor, 100 W. Capitol St., Jackson, MS 36269.

Facilities and Services: Campgrounds (some with camper disposal service); picnic areas with grills; boat launching ramp; fishing pier; swimming beach.

Special Attractions: Chewalla and Puskus lakes; reconstructed Indian mound.

Activities: Camping, picnicking, hiking, hunting, fishing, boating, swimming.

Wildlife: Threatened red-cockaded woodpecker.

UNLIKE MOST NATIONAL FORESTS, a large amount of the land within the boundaries of the Holly Springs is privately owned and, therefore, the forest cover is interrupted by dwellings, old fields, and many secondary roads, all of which add interest to the forest.

The eastern segment, which lies south and east of the town of Holly Springs, has two major recreation sites. The more interesting of the two is the Chewalla Lake region, where a variety of recreational activities can be pursued. The focal point of the area is the elongated Chewalla Lake. The day we were there, a stiff breeze was blowing across the turbid water. Under less harsh conditions, the lake is a fine one for boating, fishing, and swimming. There is a boat launching area, a fishing pier that juts into the water, and a swimming beach with bathhouse. A nature trail begins at the boat launching area and borders the lake for some distance. It passes heavily scented pines, sweet gums, red maples, hickories, and flowering dogwoods. Rustic bridges help to keep you from getting your feet wet in the streamlets that drain into the lake.

Since the 260-acre Chewalla Lake occupies the site of an earlier Indian civilization, there is a small reconstructed Indian mound protected by a split rail fence that is worth observing. Thirty feet away is a wooden observation deck that provides a sweeping view of the lake. From the deck can be seen outcroppings of red sandstone along the lake shore. The oak-leaved hydrangea, with its jagged-edged leaves, grows from crevices in the sandstone. There is a secluded campsite upslope from the Indian mound.

Ten miles east of Oxford, on the banks of Puskus Lake, is another pleasant recreation site where we camped, picnicked, and hiked in a mixed woodland of pines and hard-

woods. Boating and fishing are good in the 91-acre lake.
Hunters will like the Upper Sardis Wildlife Management
Area near here. It has a number of primitive campsites.

Three miles east of Interstate 55, in the western unit of
Holly Springs, is Tillatoba Lake, a 40-acre impoundment
with a small beach and a boat launching area. There is a
campsite here, also.

Homochitto National Forest

Location: 189,000 acres in southwestern Mississippi,
southeast of Natchez between Meadville and Gloster. U.S.
Highway 84 and State Route 33 are the major access roads.
Maps and brochures available at the district ranger stations
in Meadville and Gloster or by writing the Forest Supervisor,
100 W. Capitol St., Jackson, MS 36269.

Facilities and Services: Campgrounds (Clear Springs
has camper disposal facility); picnic areas with grills; swimming beach.

Special Attractions: Subtropical vegetation; Pipe's Lake.

Activities: Camping, picnicking, hiking, hunting, fishing,
swimming, boating.

Wildlife: Threatened red-cockaded woodpecker and American alligator.

YOU HAVE NOT EXPERIENCED the subtropical world of the
Homochitto National Forest unless you have been to Pipe's
Lake, a wonderfully secluded spot in the extreme southwestern corner of Mississippi. Reached via eight miles of
narrow graveled road west from Mississippi Route 33, spring-
fed Pipe's Lake is nestled in a ravine surrounded by forested
ridges with a flora that is mostly subtropical. Among the
plants you will see are saw palmettos scattered in the ravine
bottoms. Wispy gray strands of Spanish moss, another subtropical touch, hang from many of the trees. Also abundant
are huge, shiny- and leathery-leaved southern magnolias.
Several of them tower majestically above the ridgetop picnic
area. Red maples are also plentiful and provide brilliant red-
winged fruits in April and red-tinted leaves in October.
Added to these are temperate species such as American

hollies, pines, wild black cherries, Spanish oaks, and many other trees. A shrub known as horse sugar, near the picnic grounds parking area, has small spheres of delicate yellow-white flowers in the early spring. Keep your eyes open for the uniquely southern cross-vine that crawls up the trunks of some of the trees. It can be recognized by its several pairs of shiny, lance-shaped leaves. Floating on the water surface in some of the shallow necks of Pipe's Lake are the oval-shaped leaves of the water shield, a plant related to more colorful water lilies. There is a primitive hiking trail near the lake.

Most visitors to the Homochitto use the Clear Springs Recreation Area for their base, and for good reason. It is one of great natural beauty. The area is set among densely forested hills with steep slopes and deep ravines. In one deep, wet ravine just behind the picnic area are colonies of the giant cinnamon and royal ferns. Towering overhead are fragrant pines and white-flowered southern magnolias. Flowering white dogwood and big-leaf magnolia are also present. The huge, mottled, round leaves of the purple trillium add further color to the early spring scene.

The 12-acre lake at Clear Springs has a swimming beach and bathhouse. An interesting hiking trail completely surrounds the lake. It is a little over a mile long and is relatively easy except after a rain, when certain spots become slippery and treacherous.

Tombigbee National Forest

Location: 66,000 acres in east-central Mississippi, comprised of a unit east of Houston and a unit southeast of Ackerman. The Natchez Trace Parkway passes through both segments of the forest, as does State Route 15. Maps and brochures available at the district ranger station in Ackerman or by writing the Forest Supervisor, 100 W. Capitol St., Jackson, MS 36269.

Facilities and Services: Campgrounds with camper disposal service; picnic areas with grills; swimming beach; boat launching areas; fishing piers.

Special Attractions: Davis and Choctaw lakes; Owl Creek Archeological Site; restored Indian mounds.

Activities: Camping, picnicking, hiking, fishing, hunting, boating, swimming.

Wildlife: Threatened bald eagle and red-cockaded woodpecker.

THE TOMBIGBEE NATIONAL FOREST is divided into two parts that lie on either side of the Natchez Trace Parkway, a route that crosses Mississippi diagonally for nearly 300 miles.

Recreation in the upper segment of the Tombigbee east of Old Houlma centers around Davis Lake, a 300-acre body of water surrounded by pines and hardwoods. There is a lakeside campground beneath tall pines that is extremely attractive, particularly when the gentle breezes whisper through the needle-covered branches. Nearby is a boat launching area and a small beach. A short way out in the water is a diving platform.

A gravel road along the side of the lake across from the campground is excellent for hiking or driving. From this road, hiking forays can be taken into several inviting ravines or along dry, wooded ridges. Tulip poplars and sugar maples provide shade in the ravines where a fine array of spring wildflowers occurs. Oaks and hickories dominate the ridges. In mid-April, when the flowering dogwood blooms and the other trees are unfolding their young, fresh leaves, these forests are extremely attractive.

A short distance north of Davis Lake is a significant archeological site where five flat-topped Indian mounds apparently dating back to between A.D. 300 and 1500 were located at one time. Several artifacts have been obtained from this site, which is known as the Owl Creek Archeological Site. Two reconstructed mounds at the edge of a bottomland forest now mark the site. Wooden steps have been provided for access to the top of the larger mound. The mounds probably represent a small ceremonial center, and the larger one may have supported a temple.

Choctaw Lake is in the southern unit of the Tombigbee near Ackerman. In addition to the usual camping, picnicking, boating, and swimming facilities, there is an easy nature trail that brings you close to the trees and wildflowers that abound in the forest.

Chattahoochee National Forest

Location: 747,000 acres in northern Georgia in two sep-
arate units on either side of Dalton. Interstate 75 and U.S.
Highways 19, 76, and 441 provide access to the forest.
Maps and brochures available at the district ranger stations
in LaFayette, Blairsville, Clarkesville, Dahlonega, Chats-
worth, Clayton, and Blue Ridge, at the Brasstown Bald
Visitor's Center, or by writing the Forest Supervisor, 601
Broad St. SE, Gainesville, GA 30501.

Facilities and Services: Campgrounds (some with camper
disposal service); picnic areas with grills; visitor's center at
Brasstown Bald; boat launching areas; swimming beaches.

Special Attractions: Many waterfalls; Chattooga Wild
and Scenic River; Ellicott Rock Wilderness.

Activities: Camping, picnicking, hiking, backpacking,
hunting, fishing, swimming, boating, nature study.

Wildlife: Endangered and threatened red-cockaded woodpecker, bald eagle, and American peregrine falcon; bear,
bobcat.

EARLY IN YOUR VISIT to the Chattahoochee National Forest,
make it a point to go to the Brasstown Bald Visitor's Center
perched on top of Georgia's highest mountain at 4,784 feet.
This modernistic center is comprised of a low circular building that houses the information desk, a native crafts exhibit,
and other displays of the forest, and a tall building that
contains a theater, an observation deck, and a lookout tower.
From the tower you get an unobstructed view of the southern
Appalachians. The mountain laurels that live at the high
elevation on Brasstown Bald bloom in June, 21 days after
this species flowers in the valleys. From the visitor's center
you can chart your visit to the forest.

We began by seeking out some of the many waterfalls
and rushing mountain streams for which the Chattahoochee
is noted. First on our list was Anna Ruby Falls, a unique
double waterfall formed by two separate creeks, Curtis and
York, both of which originate up on Tray Mountain and
cascade for 153 and 50 feet, respectively, at Anna Ruby
Falls. A half-mile paved trail crosses a rushing stream twice

on its climb to the double falls. Paralleling the trail is typical dense Appalachian vegetation composed of magnolias, rhododendrons, and mountain laurels above an understory of partridge berry, orange-fringed orchis, and the green-and-white-leaved rattlesnake plantain orchid. Do not expect solitude here, however, because the double falls are extremely popular. There are picnic tables available.

Our next waterfall visit took us to the DeSoto Falls Scenic Area, a beautiful part of the forest along U.S. Route 19. There are actually three falls here within three miles of each other. A trail that parallels Frogtown Creek and connects the falls passes through lush vegetation of hemlock, white pine, and rhododendron. Mottled-leaved wintergreen, the delicate lady fern, and the curious nongreen flowering plant that looks like a pipe and is called Indian pipe are plentiful. The Upper Falls can be seen from the major highway.

The next falls we visited was Raven Cliff Falls, just off the Richard Russell Scenic Highway. This falls dramatically splits a solid rock face nearly 150 feet perpendicular from the ground. Behind the split rock the water drops approximately 75 feet before rushing through the split rock face and dropping nearly 40 feet into a deep pool. From this pool the water cascades to the main stream another 20 feet below.

We then went on to the High Shoals Scenic Area where there are enough waterfalls to satisfy the most ardent cascade lover. There are five of them in a distance of 300 feet. The second one drops 100 feet into Blue Hole, a pool of clear blue water. Below the falls is a gorgeous stand of white pine.

Going over to the western unit of the Chattahoochee, between Dalton and Rome, we first visited the Keown Falls Scenic Area. Here John's Mountain, a narrow, sharp-topped, 1,000-foot-high ridge, drops rapidly at a rock cliff on the eastern side. There are waterfalls on two streams found in the bowl-shaped escarpment on John's Mountain. Below and around the bottom of the falls lie large boulders. On the upper slopes and dry ridges are Virginia and shortleaf pine. Elsewhere the vegetation is dominated by oak, hickory, beech, and tulip poplar.

In addition to all the pounding forest cascades, you will find that babbling mountain streams are everywhere and are always clear and beautiful. Two attractive ones we ran across were Coleman River and Cooper Creek. Both of these are lined with flowering dogwood, rhododendron, mountain laurel, white pine, hemlock, and a couple of kinds of azaleas. The wildest river of all, the Chattooga, a designated Wild and Scenic River, forms the boundary between Georgia and South Carolina. This scenic, often turbulent, whitewater river is described more fully under the Sumter National Forest entry.

We became aware of how small man is when we walked among the giant trees at Sosebee Cove. This virgin forest enclave contains some of the largest hardwoods in the eastern United States. The largest Ohio buckeye we have ever seen is here, and so are huge tulip poplars well over 100 feet tall and 17½ feet in circumference. The forest floor is laden with ferns and showy wildflowers, including solid acres of Dutchman's-breeches. When we emerged from these woods, we had a deeper respect for nature.

Several interesting archeological sites are in the Chattahoochee. One is Track Rock. It is located in a low narrow gap between Blue Rock Mountain and Buzzard Roost Ridge, and is marked by four large soapstone boulders. Another site is the Blood Mountain Archeological Area. It is a 28-acre region of jumbled rocks near the site of a critical Creek and Cherokee Indian battle.

The Appalachian Trail winds for 83 miles in the Chattahoochee National Forest, staying mostly at the higher elevations. Some of the ascents along this section of the trail are long and arduous, so make sure you know your physical limitations before starting. The southern terminus of the Appalachian Trail is in the Chattahoochee. A very small part of the Ellicott Rock Wilderness is in this forest; it is described in the Sumter National Forest entry.

There are several large lakes where you can swim or boat, including Blue Ridge Lake, Lake Russell, Rabun Lake, Conasauga Lake, and Lake Chatuge. Lake Winfield Scott, located in an appealing mountain setting, has a nice swimming area. Nearby is the Buffalo Nut Trail, named for a usually rare shrub that is common along this trail. You might be more impressed by the unbelievable amount of lady fern along the trail.

If you want to camp near a lot of activity, use the campgrounds at most of the lakes. If it is seclusion you are after, let us suggest the small, densely shaded campground at Andrews Cove, complete with a mountain stream running through it.

Oconee National Forest

Location: 109,000 acres in two units in central Georgia on either side of Interstate 20 between Athens and Macon. Maps and brochures available at the district ranger station in Monticello or by writing the Forest Supervisor, 601 Broad St. SW, Gainesville, GA 30501.

Facilities and Services: Campgrounds (camper disposal service at Lake Sinclair and Oconee River); picnic areas with grills; boat ramps; swimming beaches.

Special Attractions: Scull Shoals archeological and historical areas.

Activities: Camping, picnicking, hiking, hunting, fishing, swimming, boating, canoeing.

Wildlife: Threatened red-cockaded woodpecker; black bear.

THE OCONEE NATIONAL FOREST sprawls over a part of the rolling hills of the Georgia piedmont. It does not contain large continuous blocks of public land. Instead, private property is interspersed throughout the area. When you hike or hunt, be sure to respect the private property.·

One place you will want to visit is the Oconee River Recreation Area. The entrance road leads through a fragrant grove of pines and puts you in the right frame of mind for an outing. The campground is equally conducive. The Oconee River is suitable for canoeing, although when the water level is low, there are problems with logs and sandbars. The river passes by some of the choice woodlands in the forest.

Visitors should also not miss the one-mile hiking trail from the Oconee River Recreation Area north to see the site of bridges and other remains of the famous and historic old village of Scull Shoals. The town was settled in 1784, became a fort, and boasted Georgia's first paper mill in 1812 and its first cotton gin in 1834. Even one of Georgia's governors grew up in Scull Shoals. But the village declined after the Civil War.

Another archeological site two miles north of Scull Shoals contains two prehistoric Indian mounds. Since the area is rather remote and fragile, check with the district ranger near Greensboro for the current conditions.

In that part of the Oconee west of Eatonton, you enter Uncle Remus country. Author Joel Chandler Harris lived and worked in this area. It does not take much imagination today to put yourself in Br'er Rabbit's place in the forest.

Hunting and fishing are the most popular activities in the Oconee National Forest. The 27,000-acre Cedar Creek Game Management Area, used by permit only, has a large population of white-tailed deer. Wild turkey, quail, and much waterfowl attract hunters to the Oconee. The waters of Lake Sinclair and the Ocmulgee River seem to be the best for fishing, although some of the small and even unnamed ponds of the forest are often productive.

Apalachicola National Forest

Location: 559,000 acres in the Florida panhandle between the Apalachicola River and Tallahassee. Interstate 10 is near the northern end of the forest; State Routes 20, 65, 67, 267, and 365 serve the forest. Maps and brochures available at the district ranger stations in Bristol and Crawfordville or by writing the Forest Supervisor, 2586 Seagate Drive, Tallahassee, FL 13549.

Facilities and Services: Campgrounds (some with camper disposal service); picnic areas with grills; swimming beach; campground for the handicapped at Trout Pond.

Special Attractions: Leon Sinks, unusual sinkhole ponds; Bradwell Bay Wilderness.

Activities: Camping, picnicking, hiking, hunting, fishing, swimming, canoeing, nature study.

Wildlife: Threatened red-cockaded woodpecker and American alligator.

THE MOST DISTINCTIVE FEATURE of the Apalachicola is a series of limestone sinks or depressions southwest of Tallahassee near the Leon-Wakulla county line. Known as the Leon Sinks, these depressions have taken various shapes. Dismal Sink is the most outstanding; it drops 75 feet straight down to a deep pool. Sheer rock walls rise dramatically about 20 feet above the water. The sink is 200 feet across at the top. The drier soil around the top of the sink supports such plant life as persimmons, pignut hickories, mockernut hickories, and red oaks. In moister areas below the sink are laurel oaks, white oaks, water oaks, southern magnolias, basswoods, and sweet bays. The other sinks in the area have equally intriguing names—Black, Hammock, Gopher, and Natural Bridge. A visit to the sink area is a must.

A casual drive through the Apalachicola National Forest itself may seem on the surface similar to driving through one large, homogeneous pine woodland. But if you get off the beaten path, you will have a chance to see many different plant habitats. Much of the forest is fairly uniform, with high pinelands dominated by longleaf pine and turkey oak and flatwoods dominated by longleaf pine and palmetto. Near Florida Route 65 five miles south of Wilma is a unique stand of nearly 100-foot-tall slash pines towering over sweet-bay magnolia, sweet gum, swamp bay, southern magnolia, water oak, and the eye-catching Atlantic white cedar. Along State Route 267 where it connects with Forest Road 354S is a sweet-bay swamp where sweet-bay magnolia, with its handsome, smooth-edged leaves, is joined by choice specimens of swamp bay and loblolly bay. At another sweet-bay swamp along the New River northeast of Sumatra, the Atlantic white cedar is an added species.

Where Florida Route 13 crosses the Sopchoppy River, visitors can find a good example of a creek swamp where swampy conditions prevail on either side of the creek. An interesting tree in this swamp is called the Ogeeche lime, which is actually not a lime but a type of black gum. Other trees in the creek swamp include loblolly pine, bald cypress, red maple, tulip poplar, water oak, and sweet gum.

Look for the Rocky Bluff Scenic Area on the east bank of the Ochlockonee River where unusual 15-foot sheer vertical cliffs rise above the river. These bluffs are surrounded by a rich, southern mixed-hardwood forest. You will see the southern magnolia, with its glossy, leathery leaves, as well as flowering dogwoods and redbuds. Other plants here are the loblolly pine, white ash, diamond-leaved oak, bluff oak, sweet gum, and tulip poplar. At one place in the forest is the only location in the world for the endangered Harper's beauty, a member of the lily family.

The other special attraction in the Apalachicola is known as the Bradwell Bay Wilderness. It is a 23,432-acre roadless area northwest of the community of Sopchoppy. Although this region appears to be flat, slight depressions and low ridges provide enough differences to support several different kinds of plant communities. A huge segment of Bradwell Bay is called a titi swamp, where three kinds of woody plants known as titi bushes dominate. Titi bushes have small, toothless leaves that grow on their multibranched stems. Azaleas and the odorless bayberry are also in the wilderness, as are pond pines, hardwood swamps, and even grassy areas. The wilderness is normally wet with shallow standing water, so be prepared to get your feet wet if you decide to hike here. In wet areas along some of the access roads to Bradwell Bay, look for pitcher plants, sundews, and pinguiculas, all plants that trap insects; a creeping ground pine, and a low, shrubby, yellow-flowered Saint-John's-wort. The Florida black bear is in the wilderness, as well as the threatened red-cockaded woodpecker and American alligator. We did not see a bear, but we did spot a basking alligator and heard what we thought was a red-cockaded woodpecker.

There are many activities possible in the forest besides nature study. The Ochlockonee River Canoe Trail extends for 67 miles through the Apalachicola, and the scenery along the river is superb and thrilling. So is the fishing for black bass, bream, shellcrackers, speckled perch, and catfish. In periods of low water, however, numerous sandbars and snags make canoeing more difficult.

Although there are several campgrounds in the forest, the one at Silver Lake is probably the finest. Among amenities are a 250-foot white sand beach and a nature trail. Cypresses reflected in the shimmering water of Silver Lake add atmosphere to this site.

One most special feature of the Apalachicola is its Trout Pond Recreation Area for the Handicapped. Open during daylight hours from April 1 to October 31, this area can accommodate many different types of handicapped persons. Among accommodations are 6-foot-wide paved trails; a 32-

foot-long, railed fishing dock; recorded messages operated by push buttons; and specially designed playgrounds, picnic tables, and rest room facilities. Reservations for use of this area can be made through the district ranger's station in Crawfordville.

Ocala National Forest

Location: 382,000 acres in central Florida, three miles east of Ocala. Florida Routes 40 and 19 intersect near the middle of the forest. Maps and brochures available at the district ranger stations in Ocala and Eustis or by writing the Forest Supervisor, 2586 Seagate Drive, Tallahassee, FL 13549.

Facilities and Services: Campgrounds (some with camper disposal service); picnic areas with grills; boat ramps; concessions; canoe rentals; visitor's center at Juniper Springs.

Special Attractions: Only entirely subtropical forest in the United States; Oklawaha Wild and Scenic River.

Activities: Camping, picnicking, hiking, hunting, fishing, boating, canoeing, swimming, snorkeling, scuba diving.

Wildlife: Threatened red-cockaded woodpecker, bald eagle, and American alligator; black bear, bobcat.

THE OCALA is the only entirely subtropical national forest in the United States. If you want to experience these tropics fully, take your own boat or canoe (or rent one from among the several concessionaires) and head down the black, shining water of haunting Get Out Creek. The creek is so shallow that you can nearly see bottom. An occasional school of bream may dart past and disappear beneath the always encroaching water hyacinths, or an American alligator may be seen basking along the shoreline.

As you proceed downstream, you may have to duck the Spanish-moss-draped limb of a live oak, which tried years ago to fall into the water but was prevented from doing so by a dense entanglement of catbriers. An occasional nest of a rare osprey may be seen overhead. Tall cabbage palms are so thick at times that they line both sides of the narrowing creek. Your excitement will grow with the appearance of each new kind of bird. You may be lucky enough to see an awkward-looking and very rare limpkin as it swoops away from the bank where it has been interrupted from its meal of apple-shell snails.

If you are canoeing, you can continue in the jungle corridor that lies ahead. If you are motorboating, you will eventually have to turn around before the whirring motor of your boat scrapes the bottom of the increasingly shallow creek. Except for an occasional fisherman who may be angling over the side of his nearly stationary boat for bass or shellcrackers, you will have the feeling of complete solitude.

If you prefer to do your canoeing with many fellow canoeists around, try the popular canoe run along Juniper Creek from Fern Hammock at the Juniper Springs Recreation Area to Sweetwater Springs on Florida Route 19. Along this splendid canoe course are bears, deer, American alligators, otters, herons, egrets, purple gallinules, and many species of waterfowl. Juniper Springs itself emits eight million gallons of water each day and is in a beautiful woodland setting. Near the springs is a swimming pool with an adjacent bathhouse, a concessionaire who rents canoes and sells some groceries, a self-service laundry facility, and an extensive campground.

Alexander Springs, the largest freshwater springs on federal land, is located in the Ocala. It boils forth 78 million gallons of 74-degree water each day from a large limestone cavern 25 feet below the surface. Swimming, boating, scuba diving, and snorkeling are popular activities here. The well-kept Tropicana Nature Trail skirts part of the Alexander Springs area and passes through a subtropical paradise where bald cypress, cabbage palm, saw palmetto, the dwarf palm-like coontie, and sweet bay may be examined at close range. Boardwalks enable you to take the trail without getting your feet wet. There is a concession stand and a bathhouse between the springs and the campground.

The first impression that a visitor has upon entering the Ocala National Forest is of a huge sea of pine-dominated sand. Known as the "Big Scrub," this sandy habitat comprises more than half of the forest. The Big Scrub, which contains the largest stand of sand pine in the country, is intermixed with the dwarf live oak, the sand live oak, and several evergreen shrubs. A vast deer herd roams the Big Scrub, and black bears are occasionally seen as well. Within the Big Scrub are some old sinkholes, including Blue Sink, a nearly circular pond about 100 feet in diameter and approximately 45 feet deep, filled with azure blue water. Throughout the Big Scrub, on slightly higher elevations, are extensive stands, called "islands," of longleaf pine. In the older stands of longleaf pine may be found the threatened red-cockaded woodpecker.

Of historic interest are Pats Island and Hughes Island, each covering about 900 acres. Settlers came to these islands by 1840; the life of the pioneers is vividly depicted in *The Yearling*, the Pulitzer Prize–winning novel of 1938 by Marjorie Kinnan Rawlings. Mrs. Rawlings lived with the Cal Long family on Pats Island when she wrote the book. Members of the Long family, prominent characters in *The Yearling*, are buried in the small Long Cemetery on Pats Island. The largest turkey oak in the world is also on this island.

Hughes Island, approximately four miles west of Pats Island, is also mentioned in *The Yearling*. On this island is an Indian burial mound as well as one of the largest loblolly bay trees on record.

The Oklawaha River, designated one of the nation's Wild and Scenic rivers, forms the northwestern and northern perimeter of the forest. Much of the Oklawaha has retained its original condition, particularly the nine-mile section from Rodman Dam to the St. John's River.

More than 600 lakes perforate the Ocala, ranging in size from less than an acre to Lake Ker's 2,700 acres. Each lake has its own flora attractions. Lake Ker boasts a hardwood hammock peninsula that extends into the lake. Lake Dorr, at the northwestern border of Blackwater Swamp, has a 2,500-acre tract of swamp hardwoods. One 220-acre stand of old growth cypress and tupelo gum is probably the largest in the forest. Lake Charles, a 600-acre circular natural lake, offers a ring of pond cypress containing many osprey nests.

Almost everyone can find something of interest in the Ocala. Archeologists try to piece together the ways of life of the earliest Indians by studying the many shell middens found in the forest, such as those on Kimball Island and on Bower's Bluff. Ornithologists go to the Church Lake Prairie and Mud Prairie Lake to observe the greater and the Florida sandhill cranes. History buffs are fascinated by Cat Head Pond where legend has it that near the end of the Civil War, with Union troops forging north up the St. John's River, Confederate soldiers manning Fort Gates rolled their cannons one and a half miles southwest to Cat Head Pond and buried the cannons in the water.

The Ocala Trail, inundated in spots following heavy rains, winds for 66 miles past natural ponds, through swamps of bald cypress and tupelo gum, and over the drier islands of longleaf pine. Boardwalks elevate the hiker above some of the swamps. Mosquito repellent is a must during the summer and autumn.

Osceola National Forest

Location: 157,000 acres in northern Florida northeast of Lake City. Interstate 10 passes through the forest. Maps and brochures available at the district ranger station in Lake City or by writing the Forest Supervisor, 2586 Seagate Drive, Tallahassee, FL 13549.

Facilities and Services: Campgrounds (Ocean Pond is developed; the remainder are primitive); picnic ground with grills at Olustee Beach; boat ramps; swimming beach.

Special Attractions: Ocean Pond with bald cypresses.

Activities: Camping, picnicking, hiking, hunting, fishing, boating, canoeing, swimming.

Wildlife: Threatened red-cockaded woodpecker and American alligator; black bear.

THE FOCAL POINT of the Osceola National Forest is the nearly circular 1,760-acre lake known as Ocean Pond. Swollen-based bald cypresses, with delicately draped Spanish moss hanging from their branches, grow in the shallow water of the lake. Cypress knees are plentiful, projecting above the water like wooden cones and enabling these trees to exchange gases while still growing in water. It is a magnificent sight at dawn when you can see the cypresses shrouded in the early mist. Ocean Pond has excellent camping facilities and a boat ramp is available near the campground. Fishing for bass and bream is good either from the shoreline or from a boat. Along the southern shore is the Olustee Beach Recreation Area, where one also can swim and picnic. A boat launch area is there, as well.

In other areas of the Osceola are numerous depressions filled with forests of hardwood swamps. The swamps are dominated by bald cypress, tupelo gum, sweet bay, and red maple. On drier, elevated land, stands of slash pine and longleaf pine occur. Much of the Osceola is designed for primitive camping.

The Osceola Trail, a segment of the Ocala Trail that penetrates much of Florida, passes through a part of the Osceola. Hikers and botanists will find it of interest since it goes through a variety of habitats, including hardwood swamps, but be prepared for different kinds of conditions. Wet areas will be encountered, and mosquitoes are ever present.

There is a fairly delightful canoe trail beginning at the East Tower Primitive Campsite along Florida Route 250. It follows the Middle Prong Stream until it junctions with the St. Mary's River outside of the forest boundary. Nearly 5 miles of the 11-mile trail are in the Osceola National Forest.

Croatan National Forest

Location: 157,000 acres along North Carolina coast, near New Bern and Morehead City. U.S. Highway 70 bisects the forest. Maps and brochures available at the district ranger station in New Bern or by writing the Forest Supervisor, 50 South French Broad Avenue, Asheville, NC 28802.

Facilities and Services: Campgrounds (camper disposal service at Neuse River); drinking water at Neuse River and Cedar Point; picnic grounds with grills.

Special Attractions: Largest collection of insect-eating plants in any national forest; native stand of typical mid-western trees at Island Creek Forest; world famous raised bogs; superior bird-watching; saltwater fishing along Neuse River.

Activities: Camping, picnicking, hiking, hunting, fishing, boating, canoeing, swimming, nature study.

Wildlife: Endangered and threatened bald eagle, American peregrine falcon, eastern brown pelican, red-cockaded woodpecker, American alligator, Atlantic sturgeon.

CROATAN NATIONAL FOREST has the greatest collection of insect-eating plants in any national forest in the United States. You will discover at least 11 different species of these curious plants and the special habitat each requires to survive. There are three kinds of sundews, four bladderworts, two pitcher plants, and the Venus's-flytrap. Some require spongy growing conditions; others live in the water of the shallow ponds. All have one thing in common: they survive by trapping insects and converting them into their life food. Sundews have leaves that are covered with sticky, glistening hairs that nab small insects just as flypaper does; pitcher plants have long, tubular, water-filled pitchers into which unwary insects fall; Venus's-flytraps have folded leaves that snap shut at the slightest insect touch; and bladderworts produce tiny underwater bladders that entrap minute aquatic organisms. Caution: These plants are all protected; do not pick any of them.

Remarkable, also, in the Croatan are the woods along Island Creek. You will find a native stand of beech trees, sweet gums, northern red oaks, and Spanish oaks that are more typical of trees found in the Midwest. You, of course, will also find lovely lacy Spanish moss hanging from bald cypresses, bringing you back to the reality that you are in southern climes and that the Atlantic Ocean is only 20 miles away. A nice and easy walk meanders through the Island Creek woods. Hikers in this and other parts of the Croatan are advised to bring insect repellent (the insect-eating plants can't catch them all!).

Another special feature you won't want to miss is the habitat called the raised bog, or pocosin, that is scattered throughout the forest. These are some of the most outstanding raised bogs in the world, and botanists from all over come to see them. Most of the bogs can be seen along either side of the sandy Catfish Lake Road that bisects the interior of the forest. (Drivers should be careful not to stray from this road because of the deep, treacherous sand.) The rich soil of the bogs absorbs great quantities of water like a sponge and provides an acidic habitat where you will see many members of the blueberry family thrive.

Because of its wetland environment and its location along the Atlantic Flyway, the Croatan is a must for bird-watchers. Among the endangered or threatened species seen are bald eagle, American peregrine falcon, eastern brown pelican, and red-cockaded woodpecker. The uncommon osprey is also present. The endangered Atlantic sturgeon has been found in coastal waters, and an occasional American alligator can be seen basking in the swamps along the waterways.

There are more than 4,300 acres of lakes within the Croatan's boundaries, the largest being Great Lake and Catfish Lake. If fishing is of interest, you will find it poor because of the high acidity of the lakes, but anglers more than make up for this lack by saltwater fishing at the lower end of the Neuse River. The Neuse River and Pine Cliff campgrounds along the river provide vivid sunrises; Pine Cliff is further enhanced by picturesque silhouetted pines. Our favorite camping spot is at Cedar Point near the White Oak River's junction with Bogue Sound, where you can just plain think about life in the solitude of natural beauty.

Nantahala National Forest

Location: 515,000 acres in southwestern North Carolina, south and west of Waynesville. Interstate 64, U.S. Highways 23, 129, and 441 are the major access routes. Maps and brochures available at the district ranger stations in Robbinsville, Highlands, Murphy, and Franklin or by writing the Forest Supervisor, 50 South French Broad Avenue, Asheville, NC 28802.

Facilities and Services: Campgrounds (some camper disposal stations); picnic areas with grills; boat launching facilities.

Special Attractions: Joyce Kilmer Memorial Forest of virgin trees; Joyce Kilmer–Slickrock Wilderness.

Activities: Camping, picnicking, hiking, backpacking, hunting, fishing, swimming, boating, waterskiing, nature study.

Wildlife: Black bear, bobcat.

NANTAHALA! A beautiful name for a beautiful place. There are Nantahala Gorge, Nantahala River, Nantahala Lake, Nantahala Mountain, and Nantahala National Forest. The national forest occupies much of the southwestern corner of North Carolina in some of the most super-scenic areas in the southern Appalachian Mountains and includes 80 miles of Appalachian Trail.

The gem of the national forest is the 3,800-acre tract of virgin woods that has been designated the Joyce Kilmer Memorial Forest to commemorate the man who wrote the simple poem "Trees." This is the ultimate area for inspiration and quiet meditation, where ancient hemlocks, tulip poplars, basswoods, beeches, and oaks rise to lofty heights of more than 100 feet. A picnic area next to the rushing Little Santeetlah Creek is also the starting point for the trails that penetrate the virgin forest. The most popular trail is a rather short and moderately easy loop trail that circles around to a bronze plaque dedicated to Joyce Kilmer. There are several longer trails, including one that completely encircles the memorial forest and another that parallels Little Santeetlah Creek. The peripheral trail climbs above 5,000 feet to the Haoe Overlook and to the Stratton Bald area, passing through Jenkins Meadow and Horse Cove. The Joyce Kilmer Memorial Forest occupies the southeastern fourth of the Joyce Kilmer–Slickrock Wilderness, which extends across the Tennessee line into the Cherokee National Forest. There are several wilderness trails in the Nantahala. One follows the essentially treeless Stratton Bald; another follows bubbling Slickrock Creek. Bobcats, black bears, and the introduced European boar are here, as are wild turkeys and ruffed grouse, but most of these animals try to avoid hikers.

If you would rather drive in this general area, take the forest road that parallels Santeetlah Creek before climbing to Stratton Gap on the Tennessee border. Along the way you will come upon a preserved log home, the Stewart Cabin. Break for a pause in this peaceful setting. Let us warn you that the road deteriorates after reaching Stratton Gap. Unless you have a four-wheel-drive vehicle, do not continue into Tennessee.

Another unforgettable site in the Nantahala is Whitewater Falls just above the South Carolina border south of Sapphire.

In a distance of 500 feet, the Whitewater River drops 411 feet in a series of falls and rushing cascades. There is a congenial picnic area adjacent. Another impressive falls is Glen Falls, where the river drops 50 feet over a rock ledge.. This waterfall is in a particularly gorgeous setting with a vista clear across the brilliant Blue Valley.

One of our favorite places in the Nantahala is Wayah Bald. Maybe it is because the day we visited in mid-June, the orange-flowered flame azaleas were in full bloom; maybe it is because the unsurpassed view from the stone observation tower lets you see forever; maybe it is because we were able to locate an unusual fragrant white azalea. Although the forest road climbs to 5,335 feet, it is not difficult to drive. Along the way is an abandoned but preserved early ranger station.

Another area to visit is the Standing Indian Recreation Area. It is best reached from the campground along the Nantahala River. At Mooney and Big Laurel falls are a pair of splashing cascades within one mile of each other. At Rock Gap is the second largest tulip poplar in the United States. Along the way one can enjoy many vistas, including one of Standing Indian Mountain.

Recreation-minded visitors will want to center their activities around the Hanging Dog Recreation Area situated on lovely Hiwassee Lake. The lake is noted for its bass, bream, crappie, and walleye pike. A couple of trout streams are nearby. There is also boating and waterskiing on the lake.

Pisgah National Forest

Location: 494,000 acres in the southern Appalachian Mountains of North Carolina, between the Tennessee line and Brevard, North Carolina. U.S. Highways 25, 70, and 221 pass through the forest; Interstate 40 only enters the southern end of the forest. Maps and brochures available at the district ranger stations in Hot Springs, Marion, Pisgah Forest, and Burnsville, at the Cradle of Forestry Visitor's Center, or by writing the Forest Supervisor, 50 South French Broad Avenue, Asheville, NC 28802.

Facilities and Services: Campgrounds (some with camper disposal service); picnic grounds with grills; visitor's center.

Special Attractions: Roan Mountain Gardens, spectacular natural rhododendron gardens; Cradle of Foresty, site of first American school of forestry; Sliding Rock; Looking Glass Falls; Linville Gorge Wilderness; Shining Rock Wilderness.

Activities: Camping, picnicking, hiking, backpacking, hunting, fishing, swimming, nature study.

Wildlife: Black bear, golden eagle, bobcat.

THE GREATEST PROBLEM you will have in the Pisgah National Forest is deciding what to do first. If it is the second or third week in June, you have no choice but to head to Roan Mountain on the North Carolina–Tennessee border as fast as you can get there. The summit of Roan Mountain is what is known as a bald—a natural, nearly treeless area. Here the rose-colored Catawba rhododendrons grow in profusion. The Roan Mountain Gardens are among the most beautiful natural gardens in the world. Forming a perfect background for these huge rhododendrons is a sprinkling of deep green fir and spruce. The rhododendrons are usually in bloom during the second or third week in June. Check with someone in the community of Roan Mountain, Tennessee, to see when the annual rhododendron festival is scheduled. They usually pick the right weekend. Even if the rhododendrons are not in bloom, the view from the top of Roan Mountain is superb any time of year. Local residents say you can see 6,000 peaks from the summit.

If you are a couple of weeks early for the Roan Mountain show, check out the Craggy Mountains at a lower elevation along the Blue Ridge Parkway. Although the display of Catawba rhododendrons at Craggy Gardens is outside national forest confines, the Craggy Mountains across the Blue Ridge Parkway in the forest have their share. Hiking in the Craggy Mountains, visitors can see a variety of grasses, as well as leathery-leaved shrubs of the heath family.

Another high point of the Pisgah that no other national forest can boast is a well-preserved historical site where the first school of forestry in the United States was established. Known as the Cradle of Forestry in America, the site contains several of the original buildings, dating back to the turn of the century, on their original location. Millionaire George Vanderbilt, who owned the property at that time, employed Dr. Carl Schenck, a forester from the Black Forest in Germany, to start this first school of forestry. Professor Schenck's tiny office still stands, as do the commissary, a ranger's dwelling, a lodge built in Black Forest style, and a student cabin. Begin your trip at the Cradle of Forestry by stopping at the appealing visitor's center for orientation. The center is located in a surrounding of mountain laurels and azaleas known as the Pink Beds.

Another attraction in the forest is Sliding Rock, where a flat, 60-foot-long, gently sloping rock is covered by a constant sheet of flowing water and serves as a natural sliding board. Persons in swimming attire slide swiftly down the rock and drop into a pool of water where they emerge and climb back to the top for a repeat performance. Large crowds are usually on hand to watch the participants.

A short distance south of Sliding Rock, along U.S. Route 276, is Looking Glass Falls, a 30-foot-wide cascade of white, thundering water that drops 60 feet into a catch basin. To the west is Looking Glass Rock, reached by a three-mile hike that originates from the Davidson River Road. Moisture from the vegetation on top of this giant granite rock trickles down 400-foot vertical cliffs, causing them to glisten in the sunlight and give a mirrorlike effect.

The most magnificent of the two wildernesses in the Pisgah is the 7,600-acre Linville Gorge Wilderness, a 12-mile-long, steep-sided gorge carved by the Linville River. Several hiking trails enter and traverse the area, bringing into view a number of prominent rock formations to the east. You will see that the formations labeled Hawksbill, Tablerock Mountain, and the Chimneys are appropriately named. Besides the unusual geology, the diversity of plant and animal life is rewarding. Trees of particular interest and beauty are the sourwood, sour gum, cucumber magnolia, flowering dogwood, and the delightful silverbell, with charming, white, bell-shaped flowers. You might catch a glimpse of a black bear or a white-tailed deer or a ruffed grouse. Brown and rainbow trout are in the river. On a few of the granite ledges is the endangered plant known as the mountain golden heather. If you just want to look into Linville Gorge without hiking it, drive to the Wiseman Viewpoint off Forest Road 105 for a breathtaking experience. You can picnic there as well.

The other roadless wild region in the Pisgah is the 13,000-acre Shining Rock Wilderness. A dirt road off the Blue Ridge Parkway just beyond Milepost 420 takes you to a parking lot. From here, a three-mile hike in high terrain will bring you to Shining Rock, a giant monolith composed of glistening, snow-white quartz. Because much of the area is at a high elevation, plants of the Canadian Zone, such as hemlocks, firs, and spruces, are found. Stands of rhododendrons, mountain laurels, and mountain ash add to the beauty of the land. A number of challenging ravines, called coves, invite the hiker. At the eastern end of the wilderness is the wild and scenic Pigeon River.

Our favorite camping area in the Pisgah is the Carolina Hemlock Campground, secluded enough for that good feeling you get when you camp "away from it all," but close enough to a variety of recreational activities. The campgrounds are near a stand of mighty virgin Carolina hemlocks, and a well-laid-out forest walk winds among the trees. A short distance away, along the South Toe River, is a swimming beach. Fishermen can try their luck for trout all day long in the river.

A somewhat isolated, but no less scenic, part of the Pisgah National Forest lies adjacent to the east end of the Great Smoky Mountain National Park. A highlight of this section is Rocky Bluff, a developed recreation site south of Hot Springs. Hiking trails, a stream with trout and bass, and a historic cemetery are here. In fact, you can join up with the Appalachian Trail a short distance to the west. The Appalachian Trail, by the way, has a 60-mile stretch in the Pisgah where it often parallels the North Carolina–Tennessee state line.

For a sampling of the southern Appalachian Mountains, as well as a taste of the Great Smoky area without the crowds, spend some time in the Pisgah National Forest.

Uwharrie National Forest

Location: 47,000 acres in central North Carolina in the vicinity of Troy. U.S. Highway 220 and State Route 27 are the primary roads. Maps and brochures available at the district ranger station in Troy or by writing the Forest Supervisor, 50 South French Broad Avenue, Asheville, NC 28802.

Facilities and Services: Picnic units; boat ramp.

Special Attraction: Uwharrie Mountains.

Activities: Primitive camping, picnicking, hiking, hunting, fishing, boating, canoeing.

Wildlife: Threatened red-cockaded woodpecker.

THE UWHARRIE is a small national forest in the central piedmont of North Carolina. Uwharrie Mountain, an ancient, eroded range less than 1,000 feet in elevation, is the dominating topographical feature, rising above the Yadkin River. The forest is managed primarily for timber and wildlife production. Although developed recreation sites are minimal, there are several primitive campsites, particularly along the Uwharrie Trail, and a boat launching area at the cove on Badin Lake.

The Uwharrie Trail runs for 33 miles through the eastern part of the forest, crossing or paralleling several bounding streams and then climbing and descending dry, forested ridges. The trail crosses several state highways that make easy access points. Hunters utilize this trail for deer, squirrel, quail, and rabbits. The primitive Uwharrie Hunt Camp is located midway along the trail.

If you want more scenic diversity in the forest, explore the Uwharrie Wildlife Management Area between the Yadkin and Uwharrie rivers. To get a good cross section of this part of the forest, start along the Yadkin River and gradually climb to the highest ridges in the Uwharrie Mountains. Along the river, beneath river birches, black willows, and sycamores, are the wild indigo bush and the arrowwood, both little shrubs. Back from the river's edge and before you start ascending the steep ridges is usually a bottomland forest of large, fast-growing sweet gums, tulip poplars, and American elms. During the summer, these damp, shaded woods are likely to be singing with the sounds of mosquitoes.

As you climb up out of the bottoms and away from the mosquitoes, the forests become more attractive. If you happen to choose a gentle, north-facing slope to climb, look around for stately, smooth-barked beech trees, along with the more common white, red, and black oaks, and the pignut hickory. Beneath these trees grows a fine assemblage of spring wildflowers—wild geranium, wild larkspur, may-apple, foamflower, and trillium.

Once you have reached the ridgetops, you will find yourself in a dry forest where the coarse-toothed rock chestnut oak is everywhere. In the driest areas, post oak, blackjack oak, and the large-fruited but bitter mockernut hickory join in as associates. Above some of the creeks on the steepest bluffs are dense stands of mountain laurel. This beautiful tree, more commonly associated with high mountains, usually grows with three colorful shrubs—the allspice, wild hydrangea, and pink azalea.

Here and there, particularly on southwest-facing exposures, are patches of exposed rock with sparse vegetation. Several of the plants that grow here are adapted to dry conditions—the prickly pear cactus which has succulent stems to store water; the tiny, yellow-flowered pineweed which has virtually no leaves; and the woolly lip fern which is coated with a dense mat of hairs.

The rarest form of animal life in the Uwharrie is the red-cockaded woodpecker. This species, listed as threatened on the federal list of endangered and threatened species, lives in old, diseased pine trees.

Francis Marion National Forest

Location: 250,000 acres north of Charleston, South Carolina, extending to the Atlantic Ocean. U.S. Highways 17, 17A, and 52 and several state routes serve the area. Maps and brochures available at the district ranger stations in McClellanville and Moncks Corner or by writing the Forest Supervisor, 1835 Assembly St., Columbia, SC 29202.

Facilities and Services: Campgrounds (some with camper disposal service); picnic areas with grills; boat launch areas.

Special Attractions: Revolutionary and Civil War sites; Sewee Indian shell mound; Little Wambaw Swamp; Guilliard Lake Natural Area.

Activities: Camping, picnicking, hiking, hunting, fishing, boating.

Wildlife: Endangered and threatened bald eagle, American peregrine falcon, red-cockaded woodpecker, American alligator, Florida manatee, Atlantic sturgeon.

BEGINNING 17 MILES NORTH of Charleston, South Carolina, is the Francis Marion National Forest, a forest abounding in history and possessing remarkable natural features. Francis Marion, an American settler known as the "Swamp Fox," was the first to employ guerrilla warfare when he lured the British into the swamp during the Revolutionary War. Several of the swamps remain today, as do many remnants of bygone days. Only vine-covered brick walls of Biggin Church remain standing, east of Moncks Corner. The church, originally built in 1710 and a prominent structure in both the Revolutionary War and the Civil War, has been destroyed three times by fire, the last in 1886. The rubbled ruins of the old Watahan Plantation, nearly buried by vegetation beneath a gigantic live oak, are all that remain of this once beautiful plantation, the first to be built along the Santee River. Francis Marion and his men used it as a resting place during the revolution. Old bridges such as Wadboo, Wambaw, Quinby, and Videau, now replaced by modern structures, mark the sites of other revolutionary skirmishes.

A group of earthen embankments and trenches near the Santee River can be reached by a two-mile hike through dense vegetation north from the Echaw Road. Known as the Battery, these fortifications were built during the Civil War in 1863 by General Beauregard to protect the Santee River from the encroaching Union army.

Within the forest, but on private property, are such historic structures as Middleburg Plantation, the oldest standing wooden building in South Carolina, constructed in 1699, and old and beautiful churches such as Pompion Hill and St. James. The road that passes in front of the St. James Church has been traveled by such notables as George Washington, James Monroe, and Lord Cornwallis.

Another remnant of pre–Revolutionary War days is the tar pit. Tar kilns adjacent to the pits were used to extract pitch from pinewood. The pitch would then be used to repair the wooden sailing ships. One of these pits is preserved at the Tar Pit Recreation Area.

There was activity on this land long before the revolutionary days. It is estimated that as early as 4,000 years before the Civil War there was a shell mound along the Intercoastal Waterway south of Awendaw. This prehistoric mound, measuring approximately 150 feet in diameter and rising to elevations approaching 10 feet, contains loosely packed shells, bone, and pottery shards. The mound is named for the Sewee Indians who inhabited the area until 1716. Visitation to this fence-enclosed area is permitted, but visitors are warned against disturbing anything. The trail to the Sewee Mound is interesting because it passes through a forest type not common in the Francis Marion. The trees include black oak, cherry-bark oak, live oak, loblolly pine, green ash, pale hickory, redbud, red mulberry, Hercules'-club, and three types of hollies.

Much of the Francis Marion National Forest is composed of sandy ridges where longleaf, shortleaf, and loblolly pine are common. Pond pine grows in low, flat woods. In a damp woods near Shulerville is a stand of slash pine, a tree whose trunks are slashed to obtain resin.

There are two special, unique natural areas in the forest. One is Little Wambaw Swamp, which occupies nearly 3,000 acres near U.S. 17 between Awendaw and McClellanville. This swamp fills a large depression that serves as a collecting basin from the adjacent pine uplands. The depths of the Little Wambaw contain large bald cypresses and tupelo gums. Away from the deepest parts of the swamp are occasional open ponds, where many kinds of waterfowl can be seen. Pocosins, or raised bogs, are also present. They are boggy areas dominated by evergreen shrubs. Many members of the heath family grow here; there are also insect-trapping pitcher plants.

The other special area is the unspoiled region near the Santee River that is called the Guilliard Lake Natural Area. It covers almost 1,000 acres. The finger-shaped lake was apparently an old river oxbow. An old-growth bottomland hardwood forest surrounds the lake. Huge bald cypresses and tupelo gums tower over red maples, swamp cottonwoods, water locusts, water elms, green ashes, Carolina ashes, Virginia willows, storaxes, and red buckeyes. The trail along nearby Dutart Creek passes limestone outcrops, unusual for the area.

There is a rich diversity of animal life in the Francis Marion. More than 250 kinds of birds have been recorded, including the threatened bald eagle, American peregrine falcon, and red-cockaded woodpecker. The very rare and elusive Bachman's warbler was last seen in the country in 1963 in this forest. The waters of the streams and swamps have small numbers of American alligators, and the endangered Atlantic sturgeon occupies rivers near the coast. Even the strange Florida manatee, a large aquatic mammal with flippers, has been recorded from the Francis Marion.

There are several attractive recreation areas in the forest. One of them, Buck Hall, has a boat landing ramp for easy boat entry into the Intercoastal Waterway, and another, the Huger Recreation Area, has a boat access to the scenic Huger River. The river is excellent for bass and bream fishing. Several recreation areas are situated beneath huge Spanish-moss-draped trees. Elmwood is particularly beautiful for camping and picnicking. You will appreciate its tranquility.

Sumter National Forest

Location: 358,000 acres in three units in South Carolina, two in the central piedmont near Newberry and Greenwood, the other in the southern Appalachians along the Georgia border near Walhalla. Maps and brochures available at the district ranger stations in Walhalla, Edgefield, Whitmire, Greenwood, and Union or by writing the Forest Supervisor, 1835 Assembly St., Columbia, SC 29202.

Facilities and Services: Campgrounds (some with camper disposal service); picnic areas with grills; boat launching sites; swimming beaches.

Special Attractions: Ellicott Rock Wilderness; Chattooga Wild and Scenic River; Long Cane Scenic Area.

Activities: Camping, picnicking, hiking, hunting, fishing, horseback riding, swimming, boating, nature study.

Wildlife: Threatened bald eagle and red-cockaded woodpecker.

THERE ARE TWO DIFFERENT FACES of the Sumter National Forest. One face is typical Appalachian Mountain, with blue hazy mountains, clear rushing streams, and scenic gorges. The other face is typical piedmont, with rolling hills and sluggish streams. The mountainous section presses against the Georgia state line, separated only by the wild Chattooga River. Known as the Andrew Pickens district of the forest, this 78,000-acre region has everything that the southern Appalachians have to offer.

To initiate yourself into this part of the forest, head for one of its major attractions, the 3,200-acre Ellicott Rock Wilderness, a highly scenic, roadless area that barely laps over into Georgia and North Carolina. Best approach is to take the mountain road from Walhalla north to the Chattooga picnic area at the edge of the wilderness. The wilderness is named for a rock that an early surveyor designated as the point where South Carolina, Georgia, and North Carolina come together. The trail to Ellicott Rock offers some of the best hiking in the United States. It begins in a cathedral forest of pine and hemlock, where the grandeur of nature is exhibited in its towering structures. For two and a half miles the trail stays near the East Fork of the Chattooga River, through a forest with a sampling of many wildflowers. When the trail comes to a "T" at the east bank of the Chattooga River, take the right-hand fork and begin the last two miles to the historic Ellicott Rock. Wherever you go in this mountainous part of the Sumter, you will find lush vegetation. Along at least two of the creeks is the rare and lovely oconee bells, a lovely white-flowered plant being considered for inclusion on the endangered species list.

The other star forest attraction is the Chattooga Wild and Scenic River. There probably is no river in the eastern United States more spectacular than the Chattooga, nor is there any more treacherous for the canoer, kayaker, or floater. The Chattooga rises in the mountains of North Carolina and flows for about 10 miles between the Chattooga Cliffs before coming to Ellicott Rock. From here the Chattooga begins a wild, turbulent, 40-mile stretch where it forms the border between Georgia and South Carolina. In a way it is a shame that the area is so wild, because few people ever have a chance to see the giant cliffs, massive boulders, and countless waterfalls along the way. It is even difficult to hike to much of the area, and some regions are virtually inaccessible by foot. To canoe it, only a few should even try. There are a few placid areas, such as the 6 miles from Nicholson Fields to Turnhole, but there is no gorge here, and it is the only place along the Chattooga where you can see farms and fields and homes. Unless you are an experienced canoer, do not try any of the rest, since there are rapids and falls ranging up to Class 6 that are termed "Extraordinarily Difficult to Nearly Impossible." One stretch, 7 horrendous miles from the U.S. Highway 76 bridge to Tugaloo Lake, has been described in one report this way:

> Within the first mile, the river drops over nine steep rapids. At Woodall Shoals, the Chattooga twists around gigantic shoals and drops over an 8-foot falls and down twisting, turbulent rapids. The river narrows abruptly below Woodall Shoals. In the mile to Stekoa Creek, the river rushes over two dangerous cascades, a constant series of smaller, turbulent rapids, and through a narrow half-mile-long canyon enclosed by rock walls several hundred feet high. The last 3.7 miles has 48 major rapids and cascades. It flows through an impressive gorge with cliffs rising to 400 feet. Several tributaries enter by waterfalls, such as 60-foot-high Long Creek Falls.

In the two piedmont sections of the Sumter is a completely different world of rolling hills; there are several interesting areas. Along the west bank of the Broad River northeast of Newberry is a 40-acre stand of mixed hardwoods under a sharp bluff. Many of the shellbark hickories, sweet gums, basket oaks, cherry-bark oaks, tulip poplars, and white ashes have diameters of more than three feet. The Broad River Recreation Area is adjacent. An area south of Union has been designated the Fairforest Primitive Weapons Area, where hunting is allowed only with a longbow, flintlock rifle, percussion cap rifle, or muzzle-loading shotgun.

A unique region in the Sumter is the Long Cane Scenic Area southeast of Abbeville. Long Cane Creek runs through the area and is lined in places by dense stands of giant cane, a woody grass related to bamboo. In some places along the creek, the cane may reach a height of 25 feet, an almost unheard-of size for this species. The trees that provide shade for the cane include sour gum, sweet gum, water oak, hackberry, red maple, cottonwood, sycamore, and some loblolly pine. There is a trail through some of the area.

Two highly recommended horseback riding trails are in the Sumter. The Long Cane Trail is a 26-mile loop trail that penetrates the Long Cane Scenic Area and also passes by the Parsons Mountain Lake Recreation Area. The Buncombe Trail is a 30-miler over ridges and along streams southwest of Whitmire. A good place to begin this loop trail is at the Brick House Hunt Camp.

Parsons Lake is a convenient developed area from which to explore the piedmont sections of the Sumter. There is a boat launch here as well as a swimming beach.

Cherokee National Forest

Location: 625,000 acres in Appalachian Mountains of eastern Tennessee, along the North Carolina line, on either side of the Great Smoky Mountains. Interstate 40 and U.S. Highways 19, 23, 64, and 321 serve the forest. Maps and brochures available at the district ranger stations in Etowah, Greeneville, Benton, Tellico Plains, Erwin, and Elizabethton or by writing the Forest Supervisor, 2800 N. Ocoee St. NW, Cleveland, TN 37311.

Facilities and Services: Campgrounds (some with camper disposal facilities); picnic areas with grills; boat launching areas; swimming beaches.

Special Attractions: Many Appalachian Mountains and their vegetation; Joyce Kilmer–Slickrock Wilderness.

Activities: Camping, picnicking, hiking, backpacking, hunting, fishing, swimming, boating, waterskiing, horseback riding, nature study.

Wildlife: Black bear.

TENNESSEE has only one national forest, but it is a good one. Located along the Tennessee–North Carolina border on either side of the Great Smoky Mountains, the Cherokee National Forest offers the forest visitor a total Appalachian Mountain experience. The scenery is inspiring, with unusual geological formations and a dense covering of vegetation. You can hike virtually anywhere in the forest. If you love water-based recreation, several lakes, rivers, and streams provide a diversity of fishing, swimming, and boating experiences.

Scenic splendor is everywhere. Although most of the beautiful native rhododendrons on the summit of Roan Mountain are just across the North Carolina state line in the Pisgah National Forest, the highway to the summit from the community of Roan Mountain, Tennessee, is a scenic route. You may wish to stop at Carver's Gap, in Tennessee, on your way to the top. A parking lot here serves as a good trail head for a hike into the mountain.

The Unaka Mountain Scenic Area is a mountain lover's paradise. Trails from the nearby Unaka Mountain picnic area lead into a solemn corner of the Appalachians. Take time here to explore the life of a stream or climb a ridge, observe the flowers, and listen to the songbirds—you will recharge your body batteries. Keep your eyes open for the delicate, pale pink flowers of the rare climbing fumitory, or for a clump of Fraser's sedge, with its broad, strap-shaped leaves below a stalk bearing a solitary white head of petalless flowers, or the lovely purple-flowered southern mountain gentian that blooms as late as October.

Other designated scenic areas in the Cherokee include Watauga, Rock Creek, and Falls Branch. Each of these has its own unique charm. The Watauga, located above the south shore of Watauga Lake, features dense stands of hardwoods. A good base from which to explore the Watauga is the Carden Bluff campground and picnic area. The Rock Creek Scenic Area, with nearby Benton Falls a major attraction, is adjacent to the Parksville Recreation Area, where camping, fishing, picnicking, and boating can be enjoyed in and around Parksville Lake. Falls Branch not only features a 65-foot waterfall but a stand of virgin white ash, tulip poplar, sugar maple, and hemlock. Nine miles west is the 100-foot Bald River Falls, where there is a small picnic ground. At the far northern end of the forest, and just below the Virginia state line, is the remarkable Backbone Rock, a narrow 75-foot-tall vertical rocky ridge with an old road tunneled through it. A campground and picnic area are adjacent.

Another rugged, scenic gorge is located along the Hiwassee River. Known as the Hiwassee Gorge, it is home to the very rare Ruth's golden aster, a species being studied by the Fish and Wildlife Service for possible inclusion on the endangered species list.

A neat little secluded campground is at Limestone Cove on North Indian Creek. The creek is one of several in the Cherokee known for its trout fishing.

Boating, waterskiing, and other water-related recreation can be enjoyed at Parksville Lake and other areas in the forest. At Lost Corral and Old Forge, horseback riding trails have been laid out.

The Appalachian Trail passes through or along the edge of the northern unit of the forest, routed near such attractions as Roan Mountain, the Unaka Mountain Scenic Area, and several picturesque treeless areas known as balds.

A small part of the Joyce Kilmer–Slickrock Wilderness, including Wildcat Falls, is in the Cherokee National Forest; the remainder is in the Nantahala National Forest and is described in that entry. The wilderness area is densely forested.

Daniel Boone National Forest

Location: 527,000 acres in eastern Kentucky, from Morehead south to the Tennessee border. Area is served by Interstate Highways 64 and 75, the Daniel Boone Parkway, and the Mountain Parkway. Maps and brochures available at the district ranger stations in Berea, London, Morehead, Peabody, Somerset, Stanton, and Whitley City or by writing the Forest Supervisor, 100 Vaught Road, Winchester, KY 40391.

Facilities and Services: Campgrounds (some with camper disposal service); picnic grounds with grills; boat launching facilities; swimming beaches.

Special Attractions: Red River Gorge; remarkable geological formations; Beaver Creek Wilderness.

Activities: Camping, picnicking, hiking, backpacking, hunting, fishing, boating, canoeing, swimming, waterskiing, nature study.

Wildlife: Threatened red-cockaded woodpecker.

YOU CAN BE a modern-day Daniel Boone in Kentucky's national forest, named for the old "bar hunter" himself. You can walk some of the trails he walked in the same wilderness areas; you can shoot the same kinds of weapons he shot two centuries ago.

A remarkable area known as the Pioneer Weapons Hunting Area has been set aside for the hunter who uses only a flintlock rifle, a percussion cap rifle, a muzzle-loading shotgun, a longbow, or a crossbow. The hunting area boasts of wild turkey, grouse, quail, white-tailed deer, squirrel, and rabbit. A Kentucky hunting license is required; there are campgrounds around the perimeter of the area.

The most spectacular among the many features in the Daniel Boone National Forest is the Red River Gorge. The area is known for its many geological formations, but the plant life is also unique for this part of Kentucky. There are at least 20 natural stone arches in the gorge area, including Sky Bridge, Gray's Arch, Rock Bridge, and Castle Arch. Each can be reached by trails of varying lengths and difficulties. Rock formations that suggest certain shapes are aptly named Courthouse Rock, Raven Rock, Haystack Rock, and Angel Windows. A paved loop road following Routes 15, 715, and 77 passes near many of the attractions.

Another big attraction is the Beaver Creek Wilderness. It consists of 4,800 acres along Beaver Creek. The region is varied and highly scenic, with several sandstone overhangs, or rock houses. Rocky streams and waterfalls are common. Although there are no roads and no dwellings in the area, signs of earlier habitation exist in the form of abandoned house sites, abandoned coal mines, old roads grown over with vegetation, and a cemetery.

Near the Beaver Creek Wilderness are two fine regions of unsurpassed scenery. One is Natural Arch Scenic Area, a huge sandstone arch 90 feet wide and 50 feet high. To get to it, follow a woodland trail over forested ridges and around vertical cliffs. You can see the arch in the distance from the observation point near the picnic grounds. The other is Yahoo Falls Scenic Area. It features a 113-foot waterfall and another tall sandstone arch, the Yahoo Arch. (Natives pronounce these features Yea-ho.) There are several interconnecting trails to choose from, including the Lakeside Trail along Lake Cumberland, the Skyline Trail which stays on the ridgetops, and the Roaring Rocks Trail which stays in sight of Yahoo Creek. The area represents a good example of the vegetation of the region.

There are several abandoned iron furnaces where smelting was once done. They are in varying stages of disrepair. One is Clear Creek Furnace near the Pioneer Weapons Hunting Area, another is Cottage Furnace, a well-preserved furnace north of Irvine.

Another visitor's attraction is near the picturesque, oak-forested Turkey Foot Campground northeast of McKee. There a mountain stream disappears into a limestone sinkhole during dry weather.

For hikers who only want to undertake one long trail, we recommend the Sheltowee Trace National Recreation Trail which extends for many miles throughout the entire length of the forest. On it most of the different terrains and vegetation types in the Daniel Boone can be sampled.

For the water sports enthusiasts, there are boat launching facilities at several areas, including the recreation areas at Cave Run Lake and Laurel River Lake. Waterskiing, swimming, boating, and fishing are popular during the summer when these areas are crowded.

There are several rivers and streams suitable for canoeing. The Big South Fork of the Cumberland River and the upper Red River have stretches of white water.

For the fisherman, there are trout waiting in Rock Creek, muskies, walleyes, and channel catfish in the Licking River, and bass in Lake Cumberland, Cave Run Lake, and Laurel River Lake.

NORTH CENTRAL

There are 11 national forests in the north-central part of the United States. They are the Hoosier in Indiana; Wayne in Ohio; Shawnee in Illinois; Chequamegon and Nicolet in Wisconsin; Hiawatha, Huron-Manistee, and Ottawa in Michigan; Chippewa and Superior in Minnesota; and Nebraska in the state of Nebraska.

Hoosier National Forest

Location: 187,000 acres in south-central Indiana, between Bloomington and Tell City. Interstate 64, U.S. Highways 150 and 460, and Indiana Route 37 are the major highways. Maps and brochures available at the district ranger stations in Brownstown and Tell City or by writing the Forest Supervisor, 1615 J St., Bedford, IN 47421.

Facilities and Services: Campgrounds (some with camper disposal facilities); picnic grounds with grills; boat ramps; swimming beaches.

Special Attractions: Hemlock Cliffs; Hemlock Falls and Pioneer Mothers Memorial Forest, an unusual virgin forest with giant white oaks, black walnuts, tulip poplars, white ashes, and beeches.

Activities: Camping, picnicking, hiking, backpacking, fishing, hunting, swimming, boating, nature study.

Wildlife: Threatened bald eagle.

WHEN DRIVING across southern Indiana, we like to exit I-64 at St. Croix and head a few miles north to Hemlock Cliffs, a beautiful area of wooded bluffs, dense vegetation, and fascinating caves in the Hoosier National Forest. The hemlocks, some of them with trunks nearly two feet in diameter, tower above a sweeping sandstone gorge. Mountain laurels dotted among the hemlocks add to the beauty. American chestnut trees grew on the rim of the gorge at one time, but all that is left of this poor disease-decimated species are some darkened stumps. At the far end of the gorge is Hemlock Falls, where a wide band of white water dramatically plummets to a pool below after a rain. In winter the falls in their frozen state are spectacular. In spring there is a myriad of colorful wildflowers in the canyon, including trilliums, rue anemones, bloodroots, violets, toothworts, and others. Sprinkled throughout the area are overhangs and several caves that one can visit. Arrowhead Arch, one of the caves, which tunnels about 160 feet through the sandstone, has yielded important archeological material. You can see two circular depressions on a flat table rock near this arch that probably were used in grinding grain. Hemlock Cavern is a broad sandstone overhang that could provide shelter if a storm came up.

South of I-64 near Grantsburg is a spectacular, isolated area where a narrow ridge of sandstone with tall, precipitous cliffs lies within a loop of the Little Blue River. At the easternmost end of the ridge are the remains of the old Carnes Mill. This botanically diverse area has hemlock, scarlet oak, Shumard oak, tulip poplar, American and slippery elms, basswood, and black maple. Interesting shrubs to see here are low-bush blueberry, mountain laurel, and the bush honeysuckle. If you like ferns, look around for the common woodsia, ebony spleenwort, maidenhair fern, Christmas fern, Goldie's fern, marginal shield fern, glade fern, and the polypody. The Little Blue River which flows around the bluff can be fished for largemouth and smallmouth bass, goggle-eyes, and the Ohio River muskellunge.

If you would rather be a little nearer to civilization, make your way to one of the developed recreation sites at Hardin Ridge, Saddle Lake, or German Ridge. At each of these you can camp, picnic, hike, fish, and swim. You can use your boat at Hardin Ridge (in Monroe Lake) and at Saddle Lake. German Ridge is reached by a scenic drive along Indiana 68 in the picturesque Ohio River Valley east from Tell City. The three-quarter-mile Cliffside Trail on German Ridge provides for some fine vistas and a close look at pawpaws, butternuts, persimmons, black walnuts, and dogwoods.

For the more hearty hiker, the Two Lakes Trail connecting Lake Cellina with Indian Lake is recommended. This 12-mile loop trail goes from sparkling streams to forested ridges, where an occasional wild turkey may be seen. You can also drive a forest road that connects the two lakes and that offers a number of scenic views. Boat ramps are available at both lakes, and a convenient campground is at Lake Cellina. While you are at Lake Cellina, visit the historic building that once served as the Cellina stagecoach stop and later the post office.

You can have a near-wilderness experience in the almost roadless Mogan Ridge area southeast of Tipsaw Lake. A rugged 21-mile loop trail goes across several ridges and hollows.

Before leaving the Hoosier, plan to visit the Pioneer Mothers Memorial Forest near Paoli. The center of attraction in the memorial forest is an 88-acre tract of unspoiled virgin woods featuring specimens of white oak, black walnut, tulip poplar, white ash, and beech that reach at least 115 feet tall and have a trunk diameter of 40 inches or more. There is an abundance of wildflowers in the forest, but all of them are protected from picking. This virgin forest is partly enclosed by a cedar rail fence, and the fireplace and chimney of the old Cox homestead nearby provide a nostalgic touch from pioneer days.

Wayne National Forest

Location: 176,000 acres in southeastern Ohio near the Ohio River, in three separate units surrounding Athens. U.S. Highways 33 and 50 and Ohio Highways 93, 26, and 7 are the major access routes. Maps and brochures available at the district ranger stations in Athens and Ironton or by writing the Forest Supervisor, 1615 J St., Bedford, IN 47421.

Facilities and Services: Campgrounds (some with camper disposal facilities); picnic grounds with grills; boat launching areas.

Special Attractions: Old iron furnace; rugged woodlands.

Activities: Camping, picnicking, hiking, swimming, fishing, hunting, horseback riding, boating, nature study.

Wildlife: Threatened bald eagle.

RUGGED FORESTS, picturesque creeks (called runs), and remnants of a once important iron industry highlight the Wayne Forest in Ohio's southeastern hill country. The forest is divided into three nonconnecting units, all within a 50-mile radius of Athens.

Much of the Wayne is forested with a good variety of trees, dominated by oaks, hickories, sugar maple, beech, and other splendid hardwoods. There are several trails that will allow you to enjoy this natural setting. The Wildcat Hollow backpacking trail, which originates near the Burr Oak Campground, is a rugged 13-mile, round-trip trail that alternates between ridgetops and streamsides. The trail first follows Eels Run before climbing a ridge above Wildcat Hollow and passing an old one-room schoolhouse, an old farmhouse, an oil well site, and other remnants of earlier habitation. As you finally circle back to your starting point, you will descend to Cedar Run for the pleasant last leg of your hike.

Another popular area in the forest is the Lake Vesuvius Recreation Area north of the Ohio River above Ironton. The area centers around the marvelously restored Vesuvius Iron Furnace, a structural bulwark that smelted iron ore from 1833 to 1906. From here you can hike a trail or two that is suited to your strength and to the time available, from a 3-mile signposted trail to an 8-mile trail around Lake Vesuvius to the 10- and 16-mile bluegrass and Vesuvius backpack trails in more rugged topography. You might wish to boat and fish in Lake Vesuvius or take a swim at Big Bend.

For a different panorama of the forest, seek out the two areas known as Buffalo Beats and Reas Run. Buffalo Beats, near Athens, is a one-acre area where the scrubby, shiny-leaved post oak grows among a mixture of prairie plants such as the big and little bluestem grasses and the brilliant, purple-flowered blazing star. At Reas Run, near the Ohio River at the community of Wade, the terrain is a series of rather steep ridges and hills that is drained by the placid Reas Run. The chief feature of Reas Run is a 35-acre mature stand of Virginia pine on one of the ridges above the creek. Although these 55-foot-tall pines dominate the scene, look around for the tiny pipsissewa, with its white, bell-shaped flowers, growing on the forest floor. You are likely to flush out a deer or a grouse while hiking the area.

As you explore the forest, particularly in the easternmost segment, you will find a few nostalgic covered bridges that span the Little Muskingum River.

Shawnee National Forest

Location: 254,000 acres in southern Illinois between Gorham and Elizabethtown and southward. Interstates 24 and 57, U.S. Highways 34 and 51, and State Routes 1, 13, 127, and 151 pass through the area. Maps and brochures available from the district ranger stations in Elizabethtown, Jonesboro, Murphysboro, and Vienna or by writing the Forest Supervisor, Route 45 South, Harrisburg, IL 62946.

Facilities and Services: Campgrounds (some with camper disposal service); picnic grounds with grills; boat launching areas; swimming beaches.

Special Attractions: Exceptional area for plant and animal diversity at LaRue-Pine Hills; Garden of the Gods, unique rock formations; remnants of Indian life; restored iron furnace.

Activities: Camping, picnicking, hiking, hunting, fishing, boating, horseback riding, swimming, nature study.

Wildlife: Threatened bald eagle; Swainson's warbler, bobcat.

ONE OF THE EXCITING ATTRACTIONS of the Shawnee National Forest is that it may contain the most diverse area for plants in the United States. For this reason the area, called LaRue-Pine Hills, became the first area in the national forest system to be designated an Ecological Area. In a region of four square miles, botanists have recorded a total of 1,150 different kinds of flowering plants. The waters of a spring-fed swamp lap against the base of an extensive limestone bluff that in some places rises to 300 feet above the water. On top of the bluff are acidic pebbles where the shortleaf pine, pink azalea, and other acid-loving species grow. In the swamp one discovers sponge plant, red iris, and other southern species growing at their most northern range. To see the area, we recommend taking the one-lane gravel forest service road that leads east from Illinois Route 3, five miles south of Grand Tower. The road follows the base of the bluff, separating the limestone escarpment from the swamp. After a few miles the road snakes its way up to the top of the cliff where it follows the narrow ridge, offering panoramic views of the surrounding terrain. It is against the law to pick or harm any of the plants and animals in the area.

The LaRue-Pine Hills area is a small extension of the Missouri Ozarks, but the rest of the Shawnee east to the Ohio River is a totally different ecological region of sandstone cliffs and flora. From Little Grand Canyon to Belle

Smith Springs to Hayes Creek Canyon to Lusk Creek Canyon to Pounds Hollow, sandstone cliffs alternate with deep, shaded canyons. On the bluff tops one sees post and blackjack oaks, with prickly pear cactus and American aloe growing beneath these scrub trees. The canyons themselves are dominated by stately beech trees, along with sugar maples and tulip poplars. During the spring the floor of each canyon teems with a vast assortment of wildflowers. Bluebells, celandine poppies, Dutchman's-breeches, squirrel corn, trilliums, and wild geraniums are among the more striking ones. Growing beneath overhanging sandstone bluffs is the rare French's shooting star which has flowers that have backward-pointing petals. This plant is being considered by the Fish and Wildlife Service for inclusion on its endangered species list.

The delightful Rim Rock Nature Trail just west of the Pounds Hollow Recreation Area gives a good picture of this part of the Shawnee. For a while the trail stays on top of the bluff where oaks and hickories prevail; later, by means of a wooden stairwell built in a sandstone crevice, the trail descends into a canyon rich with wildflowers.

The most popular area in the Shawnee is Garden of the Gods, where the exposed sandstone has eroded to form picturesque shapes. You will find what is known as the Camel Rock, a large formation that resembles a one-humped camel, while the so-called Monkey-Face Rock has the jutting lower jaw of a huge gorilla. A perfectly flat rock known as Table Rock has been worn completely smooth, and the Devil's Smokestack stands as an isolated narrow pinnacle.

A unique forest management technique can be seen at the Greentree Reservoir. The forest service has constructed a number of ponds in an extensive bottomland forest dominated by pin oak and overcup oak. In autumn this bottomland forest is flooded from the ponds to provide a more suitable winter habitat for waterfowl. The standing water seems to promote better growth of the pin oak, as well. In spring, when the waterfowl leave, the water is allowed to drain from these bottomland woods.

Evidence of earlier Indian habitation is also found in the Shawnee. On each of nine different bluff tops is a wall of rocks piled to a height of three to five feet. It has been suggested that Indians used these rock walls to corral wild animals such as bison that used to roam the area. Indian drawings may be seen on the smooth sandstone faces of Fountain Bluff and Buffalo Rock. At Millstone Knob, a rounded bluff, archeologists have unearthed several significant Indian artifacts.

One of the gratifying places to camp is at Belle Smith Springs. From here, in a dense woodland setting, you may explore the bluffs and canyons of the area's prettiest gorge, or you can wade in the soothing waters of Hunting Branch. There is even a natural stone "bridge" along the Sentinel Rock Trail. If you just want to picnic, climb the narrow, gravel forest road to High Knob. On a clear day you can see half of southern Illinois from the summit.

South of Harrisburg, along a massive sandstone cliff, is a rock formation that strongly resembles the face of a bulbous-nosed human. The 20-foot formation from the crown to the bottom of the chin is known as Old Stone Face.

There is a restored Iron Furnace along placid Big Creek that is well worth a visit. A nature trail that borders the creek passes the furnace, which has been abandoned for more than a century.

Horseback riders are sure to enjoy the River-to-River Trail which connects the Mississippi and Ohio Rivers. Much of the trail is in the Shawnee.

Chequamegon National Forest

Location: 847,000 acres in north-central Wisconsin, beginning about 15 miles south of Lake Superior and extending in all directions from Park Falls, the forest headquarters. The forest is served by U.S. Highways 2, 8, and 63, and by Wisconsin 13, 64, 70, 77, and 182. Maps and brochures available at the district ranger stations in Park Falls, Glidden, Medford, Hayward, and Washburn or by writing the Forest Supervisor, 157 North 5th Avenue, Park Falls, WI 54552.

Facilities and Services: Campgrounds (some with camper disposal services); picnic areas with grills; swimming beaches; boat ramps; rope tows at Perkinstown winter sports area.

Special Attractions: Excellent examples of the Great North Woods; unusual pine barrens habitat; Rainbow Lake Wilderness.

Activities: Camping, picnicking, hiking, hunting, fishing, canoeing, boating, swimming, nature study, skiing, snow-shoeing, ice fishing, snowmobiling.

Wildlife: Bald eagle, black bear, bobcat.

THE CHEQUAMEGON *is* the famous Great North Woods. Heavily forested and dotted by a myriad of lakes, this national forest is a good place to see examples of wet forests, dry forests, and boreal forests. The wettest of these northern forests is called a swamp forest because it has areas of constantly standing water. Visitors will also know when they have found a swamp forest because the dominant trees are black spruce, larch (which loses its leaves in the winter), white cedar, balsam fir, and jack pine. In swamp forests one also sees neat little evergreen shrubs of the heath family, such as Labrador tea and bog rosemary, commonly growing beneath the conifers.

Also in the Chequamegon, for contrast, hikers can see an excellent example of a dry beech-maple forest. The composition of the trees is mostly sugar maple, hemlock, beech, yellow birch, and basswood—all trees that require less water to grow. Charming wildflowers, many of them of the lily family, dot the forest floor. With increasing dryness, the forests in this area become dominated by white pine, red and sugar maples, red oak, and paper birch. All these trees require even less moisture to survive. In the driest of these northern forests, jack pine, red pine, white pine, Hill's oak, and quaking aspen prevail.

Most unusual of all three forest types in the Chequamegon is the boreal forest. A boreal forest is one that consists of plants that are characteristic of Canadian woods. The trees usually found in the boreal forests are balsam fir, white spruce, white pine, white cedar, paper birch, aspen, and maple.

The trails in the Chequamegon will lead you through examples of all three kinds of forests and permit you to get a clear picture of northern forests of the United States. Short hiking trails at Pigeon Lake, Birch Grove, Eastwood, Long

Lake, and Two Lakes provide pleasant access into the areas. We like the trail adjacent to Namekagon Lake Campground which, by the way, is one of the more picturesque places to set up camp. The trail first passes between a swamp forest and a woods dominated by white cedar. After passing a number of openings created for wildlife, the trail abruptly swings along the edge of a tamarack bog where tamaracks, or larches, predominate. After crossing a creek, the trail enters a forest of hardwood trees. You have to cross the edge of another swamp forest by means of wooden planks before you come face to face with a grove of huge hemlocks. Pause for a moment in the cool shade of these majestic giants before walking the last few feet back to the starting point of the trail. If you walked the trail quietly, you undoubtedly saw and heard a variety of birds, and you may even have caught a glimpse of a white-tailed deer.

Many of the areas in the Chequamegon without trees are meadows, where a mixture of grasses, sedges, and colorful wildflowers live together. In late summer these meadows are filled with blooming asters, goldenrods, gentians, and others.

Before completing your nature study in the Chequamegon, drive to the unique Moquah Pine Barrens which lie at the edge of a big sandy plain about 15 miles south of Lake Superior. To get to the barrens, take a forest service road north out of the community of Ino for about 7 miles. The barrens are unlike anything else in the forest. The deep, sandy soil supports large stands of jack pine and shrubs such as the narrow-leaved blueberry, blackberry, sweet fern, and serviceberry. At other places in the sandy areas, quaking aspen, paper birch, red oak, bigtooth aspen, and wild black cherry may occur. The sand is so prevalent that one may wonder how trees can establish a foothold.

For a change of pace, try canoeing some of the rivers and streams that are ideal for this form of recreation. There is a good canoe run along the West Fork of the Chippewa River from Little Clam Lake to the forest boundary. This route passes by the site of an old Indian battleground. The East Fork of the Chippewa River can be canoed from Glidden, just outside the Chequamegon boundary, to Bluebell Lake. This exceptionally long tour passes through some of the prettiest spots in the forest. For a shorter trip, canoe the

South Fork of the Flambeau River from the community of Fifield to Round Lake. You can camp along the way at Smith Rapids or Fishtrap campgrounds.

You may wish to fish for trout in the Brunsweiler River, or for harbor muskie, pike, bass, and walleye in some of the other rivers and lakes. If you swim, there are several lakes with nice beaches.

In the winter plan your recreation activities around the Perkinstown winter sports area where rope tows are available to assist skiers, or go to Mt. Valhalla, where there is a ski jump that is often used by Olympic trainees. Or just strike out across the forest on your snowshoes or cross-country skis. There are designated snowmobile trails, and you can ice fish if you like.

Another special attraction of the forest is the 6,583-acre Rainbow Lake Wilderness which can be enjoyed during any season. The many lakes that are in the wilderness provide for good fishing in both summer and winter. Hiking and cross-country skiing are fabulous here. A part of the North Country National Trail cuts diagonally across the wilderness.

One special trail in the Chequamegon that passes many land forms caused when the last glacier covered Wisconsin is the Ice Age Trail. Kettle lakes, kettle holes, and ridges of coarse gravel, called eskers, can be seen along this 40-mile trail which also goes through beautiful stands of hemlock, white pine, and white birch. The trail begins from Forest Road 101 east of the Mondeaux Flowage and continues to State Route 64 north of Diamond Lake.

Nicolet National Forest

Location: 640,000 acres in northeastern Wisconsin, a few miles east of Rhinelander. Major highways are U.S. 8 and Wisconsin 32, 55, and 70. Maps and brochures available at the district ranger stations in Eagle River, Florence, Lakewood, and Laona or by writing the Forest Supervisor, 68 S. Stevens, Rhinelander, WI 54501.

Facilities and Services: Campgrounds (some with camper disposal service); picnic areas with grills; boat access areas; swimming beaches.

Special Attractions: Giant White Pine Grove; virgin hemlock; Whisker Lake Wilderness; Blackjack Springs Wilderness.

Activities: Camping, picnicking, hiking, swimming, hunting, fishing, boating, cross-country skiing, snowshoeing, ice fishing, nature study.

Wildlife: Bald eagle, black bear, bobcat.

YOU CAN WALK without hearing a sound on the pine-needle-covered path to the Mile Post 0 treaty tree which marks the boundary between Wisconsin and Michigan, and which also separates the Nicolet National Forest from the Ottawa National Forest. There are countless such places in the Nicolet, where peace and tranquility prevail and time seems to have lost its importance. Another one of these is in the Giant White Pine Grove about one mile east of Woodbury Lake. Reached by a short trail, this 30-acre grove contains a thick stand of ancient white pines, some of which measure three feet in diameter. Another spot to enjoy solitude is in the virgin hemlock-hardwood forest between McKinley and Bose lakes. In this reverent setting, very old hemlocks, sugar maples, basswoods, and yellow birches tower above a forest floor with the fallen trunks of once giant monarchs.

Eight miles east of nearby Three Lakes is a favorite haunt of ours. Here, on the western shore of Shelp Lake, are marvelous specimens of old hemlock and pine. A boardwalk across a boggy swamp forest enables one to reach the area without unduly damaging this fragile habitat.

One easy way to get a general impression of the Nicolet National Forest is to hike the easy, mile-long Butternut Lake Forest Trail. Helpful signs along the trail explain the ecology of the forest. You will walk along an "avenue" of giant 400-year-old white pines and then enter a cathedral forest of virgin hemlock. The grandeur of these trees is unsurpassed in the Nicolet. From here the trail leads through an ancient swamp dominated by tamarack, or larch. The posted sign tells you that you are entering another world, a world that had its beginning about 50,000 years ago. The trail in places is constructed of cross logs that allow visitors to cross the swamp more easily.

The Nicolet is also a wilderness lover's dream. Two small

wilderness areas of about 10 square miles each have been designated recently. One is the Whisker Lake Wilderness, at the extreme northeast corner of the forest. It encompasses Riley Creek and a number of scenic lakes where fishing is superb. Whisker Lake itself is bordered by large pines whose needles were referred to as "chin whiskers" by the old-timers. The other roadless area is Blackjack Springs Wilderness which lies northwest of Anvil Lake. A hike along Blackjack Creek or Goldigger Creek can be rewarding.

South of Anvil Lake is the Anvil Trail Area for hikers and skiers. Six miles of hiking trails and 12 miles of cross-country ski trails have been laid out, with the ski trails graded according to their difficulty. For proficient skiers there is also the hilly Devil's Run Trail. For campers there are pleasant facilities at Anvil Lake.

Another excellent trail for both the hiker and cross-country skier is the Lauterman National Recreation Trail. This is an eight-mile loop trail, starting at Perch Lake at one end and either Lost Lake or the Chipmunk Rapids campground at the opposite end. The trail passes by many lakes, including Lauterman Lake, Little Porcupine Lake, and Mud Lake.

Fishing is at its best in the Nicolet. Some lakes and rivers are better for certain kinds of fish, some for others. Check at the ranger stations for the best places to fish for muskie, walleye, trout, and bass.

If you are in the Nicolet during autumn, you will see spectacular colors offered up by the foliage of the sugar maples, aspens, and ashes.

Hiawatha National Forest

Location: 879,000 acres in two units in Michigan's Upper Peninsula, with frontage on Lakes Superior, Huron, and Michigan. I-75, U.S. 2, and State Routes 28, 94, and 123 are the major highways in the forest. Maps and brochures available at the district ranger stations in Manistique, Munising, Rapid River, St. Ignace, and Sault Ste. Marie or by writing the Forest Supervisor, 2727 N. Lincoln Road, Escanaba, MI 49829.

Facilities and Services: Campgrounds (some with camper disposal service and some for handicapped); picnic areas with grills and some fireplaces; boat launching areas; off-road vehicle areas.

Special Attractions: Mature stands of balsam fir and white cedar; historic lighthouses and charcoal furnace; rare wildflowers such as Houghton's goldenrod and the dwarf lake iris.

Activities: Camping, picnicking, canoeing, boating, hunting, fishing, swimming, nature study, snowmobiling, downhill skiing, cross-country skiing, snowshoeing, rock hunting.

Wildlife: Endangered American peregrine falcon; black bear, porcupine, bald eagle.

THE GREAT AMERICAN POET Henry Wadsworth Longfellow immortalized Lake Superior by referring to it as the shining Big-Sea-Water in "Song of Hiawatha." Today you can observe the Big-Sea-Water from that part of Michigan's Hiawatha National Forest that borders Lake Superior from the southern end of Tahquamenon Falls State Park to Monocle Lake. Along this 30-mile stretch are an attractive picnic site at Big Pine, campgrounds at Bay View, an interesting archeological area at Naomikong Point, a ski area at Iroquois Mountain, numerous scenic vistas, and ample opportunities for swimming and fishing.

The Hiawatha National Forest is in two segments in the Upper Peninsula of Michigan, uniquely situated so that it has frontage on Lakes Superior, Michigan, and Huron. Because of the frequency of snows, there is great winter recreation in the form of downhill skiing on Iroquois Mountain and Chatham Hill, cross-country skiing, snowshoeing, and snowmobiling. When the weather gets warm, canoeing becomes popular, particularly in the AuTrain and Indian rivers. The AuTrain canoe trail is a relaxing 10 miles past a variety of wetland habitats where it is easy to see waterfowl, turtles, muskrats, and an assortment of songbirds. Fishermen can get their limit of perch, walleye, rock bass, and, during the spring and autumn, steelhead and salmon, respectively. A more difficult canoe stream is the Indian River, whose 55 miles meander through hardwood forests to conifer woods to marshes and swamps. To canoe the entire length, embark from Fish Lake, about 15 miles south of Munising. There are a number of places where you can get out. The most unspoiled section of the stream is between the Indian River Campground and the 8-Mile Bridge. The visitor gets the feeling that he is the first to enter and explore this area.

The Bay de Noc-Grand Island Trail is a must for horseback riding or long-distance hiking. Bring your own horses or rent them from concessionaires. Beginning 1½ miles east of the town of Rapid River, this trail continues for 40 miles to Ackerman Lake, having followed the old forest trail used by Indians and trappers on their route from Lake Superior to Lake Michigan and back. From rapid River the trail passes through jack pine forests and sand plains abundant with blueberries to deep forests of sugar maple and beech along the northern part of the trail.

The medium-length Bruno's Run Loop Trail is another favorite hiking trail. Beginning at Pete's Lake Campground (which is a nice place to call home, incidentally), this moderately easy trail goes over rolling hills and past bays and lakes, culminating in awesome Hemlock Cathedral, a thick stand of large hemlocks, which resembles a high-ceilinged cathedral. If you take this trail in warm weather, be sure to have your insect repellent handy. For a short hike, take the half-mile loop Juniper Trail at the scenic Point Aux Chenes. This trail alternates between low ridges of sand, marshes, and small ponds. Signs along the way are helpful in describing what you are seeing.

If you want to do something quite different, we suggest the following excursion. Make your way over to the rugged Rock River Canyon a few miles northwest of the Chatham Hill Ski Area. There are actually two canyons, each about 150 feet deep and densely forested with an assortment of hardwood species. There you will find magnificent 40-foot-deep caves that have been weathered into the sandstone. The 20-foot Rock River Falls splashes into a picturesque reflecting pool below. Several species of wildlife, including the black bear, are also found in this area.

Another adventure can be had at Horseshoe Bay, a roadless area of 4,000 acres on the shore of Lake Huron. A sandy beach is at the lower part of the area, but north of it the shoreline is marshy or rocky and difficult to maneuver. But explorers will see occasional low ridges of forest dominated by the beautiful balsam fir and white cedar, and they can also view the very rare Houghton's goldenrod that grows along the shore. Eagles and ospreys are sometimes seen overhead.

Of unusual interest to the naturalist is the Squaw Creek Old Growth Natural Area. This 64-acre special area off County Road 513 south of Rapid River contains some of the largest white and red pines and hemlocks in the entire forest. Following old abandoned roads is the best way to get around in this modern-day roadless area.

There are two islands in Lake Huron that are part of the Hiawatha and are accessible only by boat. One of our favorite places is Round Island, or Nissawinagong, in the Straits of Mackinac just off the south coast of Mackinac Island. The 378-acre island is a prominent landmark because of the nostalgic lighthouse that has stood as a beacon since its construction in 1873. At the northeast corner of Round Island are limestone cliffs that rise to a height of 76 feet above the water. The island's wooded interior is composed mainly of white cedar, balsam fir, pine, spruce, and white birch. Wildflowers include the dwarf lake iris, which is under consideration by the United States Fish and Wildlife Service as an endangered species, and the delicate and very rare calypso orchid.

Another island of the forest is Government Island, a 214-acre island in the Les Cheneaux chain a short distance south of the mainland. The rocky interior of the island contains a dense stand of white cedar and balsam fir. A narrow footpath encircles the island.

Of unusual interest are the remains of a charcoal kiln known as the Bay Furnace near the Lake Superior shoreline northwest of Munising. The furnace dates to 1870. A campground is nearby.

If you are looking for a pleasant picnic spot, seek out the Point Peninsula Picnic Ground, where you can enjoy lunch in the shadow of the scenic Point Peninsula lighthouse. This 1865 structure is situated on a peninsula that juts into Lake Michigan and is open to the public. After your picnic, walk or wade along the shoreline and look for pretty rocks or interesting fossils. If you are at Peninsula Point during September or October, you might see thousands of migrating monarch butterflies that have stopped for a rest before continuing their voyage southward.

One campground worthy of note is Camp 7 on the eastern shore of Camp 7 Lake, where there are facilities for the handicapped. There are also a nature trail and boat launch facilities at Camp 7.

Huron-Manistee National Forest

Location: In two segments, the 415,000-acre Huron stretches from Lake Huron west for 75 miles to Grayling, Michigan, and the 500,000-acre Manistee extends from Lake Michigan east to Cadillac and Big Rapids. Maps and brochures available at the district ranger stations in Baldwin, Cadillac, Harrisville, Manistee, Mio, East Tawas, and White Cloud or by writing the Forest Supervisor, 421 S. Mitchell St., Cadillac, MI 49601.

Facilities and Services: Campgrounds (some with camper disposal facilities); picnic grounds with grills; boat launching areas; bathhouses and swimming beaches.

Special Attractions: Wild and Scenic rivers such as the AuSable and Pere Marquette; waterfowl observation areas; only location in the world where Kirtland's warbler nests.

Activities: Camping, picnicking, hunting, fishing, horseback riding, canoeing, swimming (Island Lake and Loon Lake have bathhouses), snowmobiling, snowshoeing, cross-country skiing, downhill skiing.

Wildlife: Endangered Kirtland's warbler and American peregrine falcon; bald eagle.

ALTHOUGH the Huron and Manistee national forests of central Michigan used to be maintained as separate forests, they are now administered as one entity. Nonetheless, each forest occupies a separate area and has its own character.

Both forests provide ideal conditions for canoeing. Either bring your own or rent from concessionaires in the area. The recently designated Wild and Scenic AuSable River offers imposing scenery along 23 miles of its length in the Huron. Although the AuSable at one time was a major waterway for transporting logs to sawmills downstream, today it is a fisherman's paradise for native brown trout.

The Manistee has the Pere Marquette River, with the segment from Marlborough to Custer wild enough to be designated a National Wild and Scenic River. Put your canoe in at the M-37 bridge south of Baldwin and begin your peaceful journey on the smooth and gentle currents of the clean, gravel-bottomed river. Both banks are crowded with the dense growth of hardwood forests. In some areas hundred-foot dirt banks rise above the river. You will see fishermen casting for brown trout, and, if you are lucky, you might glimpse a mink or a river otter or a beaver in the water. Suddenly the forest gives way on either side to an open wetland of bayous and sloughs known as Nelan's Marsh. This is an excellent place to observe waterfowl—you may see wood ducks, mallards, black ducks, canvasbacks, scaups, and redheads. You can exit from the Pere Marquette River near the forest boundary at Indian Bridge, or you can continue on for another 20 miles to Ludington.

The most distinctive feature of the Huron National Forest is that it is the only breeding habitat in the world for the blue-gray and pale yellow Kirtland's warbler, one of the endangered bird species of the United States. This bird demands a precise nesting place in young jack pines with low brushy undergrowth. Because there are probably fewer than 300 pairs of Kirtland's warblers in the world today, the United States Forest Service has set aside about 4,000 acres in the Huron that are managed exclusively for this little bird. Management includes periodic burning to insure a continuous growth of young jack pines. In addition to Kirtland's warbler, the bald eagle and endangered American peregrine falcon can be seen overhead. For those persons wishing to study the plants and animals of wetlands, a visit to the Tuttle Marsh Wildlife Area in the southeast corner of the Huron about three miles from Lake Huron is recommended.

Michigan's Shore-to-Shore Trail cuts across nearly 75 miles in the Huron, passing through forests and wetlands and over picturesque streams. At one point the trail comes near the site of a devastating forest fire near the community of Luzerne. The trail accommodates both foot hikers and horseback riders.

The Manistee has one of the most unusual recreation areas found in any national forest. It is the Lake Michigan Recreation Area, where sand is everywhere providing plenty of beach space and surfing areas along the shore. There are also a number of trails so hikers can note the plants and animals that live in this special sandy habitat. The vast Nordhagen Dunes lie immediately east of the recreation facilities.

A wildflower sanctuary in the Manistee has been established around Loda Lake north of White Cloud, where trilliums, bluebells, and marsh marigolds are among the many flowers you will enjoy. The preserve is a joint project of the Federated Garden Club of Michigan and the United States Forest Service.

Both the Huron and Manistee teem with winter activities. Several snowmobile trails penetrate the forests, and snowshoeing and cross-country skiing are popular. Downhill skiing can be enjoyed at Caberfae in the Manistee.

There are many campgrounds to choose from, including an extensive one at the Lake Michigan Recreation Area. A pleasing site is at Pines Point near Hesperia, where the rushing sounds of the water from the South Branch of the White River will soothe your soul and lull you to sleep in a short time.

Ottawa National Forest

Location: Nearly one million acres at the western end of Michigan's Upper Peninsula, between Ironwood on the west and Iron River on the southeast. U.S. Routes 2 and 45 cross in the forest, and Michigan Route 28 goes through the forest from east to west. Maps and brochures available at the district ranger stations in Bergland, Bessemer, Iron River, Kenton, Ontonagon, and Watersmeet, at the Sylvania Visitor's Center, or by writing the Forest Supervisor, U.S. 2 East, Ironwood, MI 49938.

Facilities and Services: Campgrounds (some with camper disposal service); picnic grounds with grills; Sylvania Recreation Area.

Special Attractions: Sylvania Recreation Area (for canoeing, hiking, primitive camping); Sturgeon River Gorge; scenic Potawatomi and Gorge waterfalls.

Activities: Camping, picnicking, backpacking, hiking, hunting, fishing, canoeing, boating, swimming, snowmobiling, downhill skiing, cross-country skiing, snowshoeing, ice fishing.

Wildlife: Endangered American peregrine falcon; black bear, porcupine, river otter, bald eagle.

A UNIQUE EXPERIENCE in the Ottawa is to spend some time in the Sylvania Recreation Area, where this essentially roadless region of more than 21,000 acres provides you with the opportunity for peace and solitude amid pristine beauty. Ancient glaciers have left the area pocked with outstandingly clear lakes and ponds where one can see through 30 feet of water to the lake bottom. We suggest starting your Sylvania adventure at the modern visitor's center southwest of Watersmeet. In addition to maps and brochures, you can learn about any special conditions that may prevail, and you may look at the series of exhibits that shows the past and present use of the area. There is a quarter-mile nature trail at the center to get warmed up with. Since there are no roads, you either have to canoe or hike the Sylvania. One possibility you may wish to consider if you have your own canoe (or a rental) with you is to put into the upper end of large Clark Lake and paddle to the southeast corner. Then portage for a quarter mile to the lower end of Crooked Lake. After gliding to Fox Lake where there is a campsite, one can portage a short way to Mountain Lake. You can then alternately canoe and portage to East Bear Lake, West Bear Lake, Ker Lake, High Lake, back to Crooked Lake and Clark Lake, and to your abandoned vehicle. All along most of these lakes are primitive campgrounds. If you do not have a canoe, it is a nice two-and-a-quarter-mile hike to Mallard Campground on the shore of Loon Lake. Along the way you will pass forests of pine, spruce, fir, maple, birch, and hemlock, as well as plentiful wildflowers. You can also fish for bass, lake trout, walleye, northern pike, perch, and sunfish. Occasionally a bald eagle or a loon might be seen, or a black bear might scramble away from a lake's edge.

If it's rugged scenery and more strenuous hiking you're after, head for another forest highlight—the wild Sturgeon

River at the northeast corner of the Ottawa. The Sturgeon River has cut a deep canyon which is densely forested. From the community of Sidnaw, take Forest Road 191 north. At the Sturgeon River Campground, the road crosses the Sturgeon River and climbs to the ridge that parallels the canyon on the east. There are several views across the heavily wooded Sturgeon River Canyon before the road (which is now Forest Road 193) again descends to cross the river. Take a short side trip to Silver Mountain and test your stamina by climbing the 150-plus wooden steps to get to the top of another ridge that offers a sweeping view.

From here you can take a forest road to Bob Lake, where the well-laid-out Beaver Lodge Trail meanders past a colony of active beavers. At this and other areas in the forest, you must never be without your insect repellent during the mosquito season.

For another rewarding outing, go north from Bessemer to Black River Harbor, the forest's only shoreline footage on Lake Superior. Just before reaching Black Harbor, stop at the parking lot for the Potawatomi and Gorge falls. The National Recreation Trail to these falls is barely more than half a mile long and passes through densely shaded stands of mature hemlocks and various species of hardwoods. Several wooden stairways with resting platforms are provided. The two falls are about 800 feet from each other and well worth the hike.

If you do not want to cope with primitive camping in the Sylvania Recreation Area, try the more comfortable and modern campground along Langford Lake. There is a special feeling to frying your morning bacon and watching the sun rise from across the lake.

The Ottawa National Forest is also a paradise for winter activities. Snowmobile trails are plentiful in the forest, and ski slopes dot the surrounding area. Cross-country skiing and snowshoeing are common means of getting about.

Chippewa National Forest

Location: 650,000 acres in north-central Minnesota, 100 miles northwest of Duluth, near Cass Lake. U.S. Highway 2 bisects the forest. Maps and brochures available at the district ranger stations in Blackduck, Cass Lake, Deer River,

Marcell, and Walker or by writing the Forest Supervisor, Cass Lake, MN 56633.

Facilities and Services: Campgrounds (some with camper disposal service); picnic areas with grills; boat launching areas.

Special Attractions: Bald eagle nesting areas; orchid bogs; historic CCC camp and three-story log building.

Activities: Camping, picnicking, canoeing, boating, hunting, fishing, swimming, nature study, snowmobiling, cross-country skiing, snowshoeing.

Wildlife: Endangered American peregrine falcon; osprey, bald eagle, merlin, black bear, otter.

HIKERS AND CAMPERS will want to have binoculars, bird guides, and wildflower manuals when they head for the Chippewa National Forest, for here, among forests of giant conifers and mixed hardwoods, punctuated by more than 1,300 lakes, is a wide variety of plant and animal life sure to please all those who love nature.

Forest personnel are proud that more than 10 percent of the breeding pairs of the bald eagle in the country are in the Chippewa National Forest. Several pamphlets about the eagle have been written by resident biologists of the forest. You are most likely to see a bald eagle if you visit between March and November, because this is their breeding season. Your best bet for observing them is around Lake Winni-bigoshish. Also in the forest and requiring special protection because of their rarity are the double-crested cormorant, osprey, and merlin. In all, nearly 250 kinds of birds have been recorded.

Major efforts have been made in the Chippewa to enhance its wetlands for waterfowl and wildlife. Visit Amik Lake to see what has been done. This 108-acre lake, constructed in 1971, has as much diverse plant and animal life as anywhere in the forest. Eye-catching water lilies and water smartweeds grow with the less conspicuous sedges, bulrushes, bladderworts, and coontails. Many waterfowl may be observed here, and fish such as northern pike, yellow perch, bullhead, and sunfish are common.

There are also many wildflowers in the forest, beginning with hepaticas, bloodroots, and buttercups in the early spring.

These give way to columbines, trilliums, wild gingers, and the attractive marsh marigold in late spring, and black-eyed Susans, Indian paintbrushes, and morning glories during the summer. As late as October you can find fringed and bottle gentians in bloom.

There is a very special but fragile habitat in the Chippewa where delicate wildflowers, some of them rare, grow. These are boggy areas where white cedars, larches, balsam firs, and black spruces dominate the tree layer. In the dense shade on the forest floor beneath, growing among deep beds of sphagnum or peat moss, are such things as a pitcher plant, a dwarf raspberry, the twinflower, and several tantalizingly charming wild orchids. One such bog has as many as 11 different kinds of orchids present. It is best to observe these cedar bogs from their edge, since walking into them creates a harmful disturbance to the plants, as well as being hazardous to your own safety because of the treacherous holes in the peat. Besides, the mosquitoes are big enough and abundant enough to carry you off!

Those wishing to canoe can select from three exceptional canoe tours. The Turtle River Canoe Tour extends for 16 miles from Gull Lake to Buck Lake. There are no portages on this route, and only a few nondangerous small rapids break up the otherwise serene waters. You will paddle through one of the largest and most famous wild rice beds at Big Rice Lake. Take time to fish for walleye and northern pike.

The 18-mile Inguadona Canoe Tour from Inguadona Lake almost to Leach Lake has a few more rapids and is more difficult. But you will see large beds of wild rice and forests of balsam fir, white cedar, spruce, aspen, and paper birch, and you are likely to catch a glimpse of some raccoons, beavers, minks, muskrats, a family of otters, or even a black bear.

The 18-mile canoe run along the Rice River extends from Clubhouse Lake to the village of Bigfork. Since the river sometimes becomes too shallow for canoeing in summer, you should check conditions at the Marcell Ranger Station before embarking.

Hiking can be enjoyed along many miles of trails during the summer, while cross-country skiers find ample trails for their winter excitement. The most rewarding of the long trails that pass through a variety of vegetation types and topography are the 22-mile Cut Foot Sioux Trail and the 12½-mile Simpson Creek Trail. If you just want a leg stretcher, try the short trail from the Lake Erin Picnic Area.

Autumn brings a myriad of colors to the forest. Driving or hiking is splendid this time of year. The drive from Jingo Lake to Pughole Lake is particularly rewarding, as is the one in the rolling Suomi Hills. Pick up a fall-color tour folder and map at one of the ranger stations.

Reminders of early as well as present day Indian habitation are everywhere. Many of the names in the forest are of Indian origin. There is a Chippewa burial ground where wooden coffins lie on top of the ground. Near Little Cut Foot Sioux Lake is the Turtle Effigy Mound, reached by a short, self-guided trail. This 27-foot-long mound in the shape of a turtle was probably ceremonial in nature since it contains no burial sites. On a nearby tree is the carving of a human figure with one foot missing.

Some other unique features of the Chippewa include the sight of the Mississippi River meandering gently and unimpressively across the heart of the forest, having just begun its long journey south about 20 miles west of the forest boundary. Another feature is Star Island, located in the middle of Cass Lake; Star Island is unusual in that it has its own lake, Lake Windigo. One may take a boat out to visit the island, where there are trails and a small campground.

For history buffs, there is the Rabideau Historic Site where one of the best-preserved CCC camps in the United States can be seen. For those too young to remember, the CCC, or Civilian Conservation Corps, was started by President Franklin Roosevelt to give work to unemployed youth. Constructed in 1935 for CCC Company 708, this camp probably contained 28 assorted buildings, 16 of which still stand. You will see several barracks, the officers' quarters, two mess halls, a hospital, tool house, and pump house. This CCC company performed a variety of forest tasks, including the planting in 1937 alone of more than 500,000 trees. The company's greatest accomplishment, however, was the construction of a three-story log building that today serves as headquarters for the Chippewa National Forest. Designated a National Historic Place, this 8,500-square-foot building is made of 16,000 feet of red pine logs that are notched and grooved for perfect fitting. A 50-foot fireplace composed of 265 tons of rock highlights the interior of the building. The building cost $225,000 to construct in 1935. Imagine its worth today.

By the way, if you want to camp, try Norway Beach. Its scenic position on the southeast shore of Cass Lake cannot be beat. The cool breezes that blow across the clear water will cause you to snuggle farther down in your sleeping bag.

Superior National Forest

Location: Three million acres in northern Minnesota surrounding Ely and reaching the Canadian border. U.S. Highways 53 and 61 and State Routes 1, 12, and 169 serve the forest. Maps and brochures available at the district ranger stations in Aurora, Grand Marais, Isabella, Ely, Cook, Tofte, and Virginia, at the Voyageur Visitor's Center near Ely, or by writing the Forest Supervisor, Box 338, Duluth, MN 55801.

Facilities and Services: Campgrounds (some with camper disposal service); picnic units with grills; boat launching area; visitor's center at Ely; ski tows, lifts, T-bars, and J-bars.

Special Attraction: Boundary Waters Canoe Area.

Activities: Canoeing, camping, picnicking, hiking, hunting, fishing, swimming, cross-country skiing, downhill skiing, snowshoeing, snowmobiling, ice fishing.

Wildlife: Threatened northern timber wolf; bald eagle, black bear.

THE SUPERIOR NATIONAL FOREST in northern Minnesota is well named, for it is *the* superior national forest for canoeing. One-third of the Superior has been designated the Boundary Waters Canoe Area, a roadless region that offers a special opportunity for wilderness experience. The Boundary Waters Canoe Area stretches for more than 100 miles along the Canadian border.

Be prepared for the fact that even if you have canoed elsewhere, there is a different feeling when you glide into the Superior's lakes and past its vast wilderness of spruces and pines. As you get farther and farther from civilization, you really sense that you are in a mysterious and whole new unknown world.

No matter how long your voyage will be, you must secure a permit to enter this wilderness from the district ranger stations. There are also many outfitters in the area who can furnish you with all you need (including insect repellent). Pack as lightly as possible, because you will have to portage your canoe from time to time. There are numerous rapids and falls; if you are not familiar with the water, do not run the rapids. We advise that you visit the Voyageur Visitor's Center near Ely before starting to explore the Superior by canoe.

For the truly hearty, canoe the International Boundary Route, a trip requiring 235 miles of paddling and 9 miles of portaging. Among the lakes traversed are Rose Lake, Gunflint Lake, Saganaga Lake, Crooked Lake, Lac La-Croix, and Loon Lake. The currents at the lower end of the Basswood River and at Curtain Falls are extremely treacherous. At a number of places along this route are the famous painted Indian hieroglyphics known as picture rocks. The paintings, many of which are well preserved, are in red. Allow at least 22 days to make this trip one way, and be prepared to use primitive campgrounds where the facilities are minimal.

There are, of course, many shorter canoe routes. The two-day Sea Gull–Red Rocks loop trail is one of the best. There is only a half mile of portaging to go with 23 miles of paddling. The route, which includes Sea Gull Lake, Red Rock Lake, and Saganaga Lake, is highly scenic with many beautiful islands and rugged lake shores. Fishing for lake trout and northern pike is also good in these lakes.

If you are not a canoe enthusiast and not a long-distance hiker, there are several excellent drives that allow you to see much of the forest from your vehicle. The Gunflint Trail, for example, begins at Grand Marais and proceeds for 60 miles northwest through wilderness scenery, terminating at a campground appropriately named Trail's End. There is ample opportunity to stop along the trail to walk, picnic, fish, camp, or just enjoy the lovely vistas.

Another drive is the Echo Lake Trail from Ely to Buyck.

At Fenske Lake you can enjoy a nature trail; a little farther along, you can see the Indian picture rocks at Hegman Lake. At Echo Lake, drive north to see the Vermilion Falls, and then stop at Crane Lake. The Crane Lake Narrows, with its pine-dotted rocky shores, is a place where the visitor finds himself sitting and meditating for many minutes.

For those who like outdoor winter sports, there are more than 200 miles of trails for snowmobiles, while at the Giants Ridge, Hidden Valley, and Lutsen ski areas, there is a variety of rope tows, lifts, T-bars, and J-bars.

Nebraska National Forest

Location: 245,000 acres in central and northwestern Nebraska. U.S. Highways 20 and 385 and State Routes 2 and 97 are the major access roads. Maps and brochures available from the district ranger stations in Halsey and Chadron, or by writing the Forest Supervisor, 270 Pine Street, Chadron, NB 69337.

Facilities and Services: Campgrounds with camper disposal service; picnic areas with grills; swimming pool; tennis courts.

Special Attraction: Bessey Nursery.

Activities: Camping, picnicking, hiking, hunting, fishing, swimming, tennis.

Wildlife: Endangered or threatened bald eagle and American peregrine falcon; rare northern swift fox.

THE NEBRASKA is quite different from other national forests. Two of its three segments are composed of forests entirely planted by man. These are the Bessey Unit and the Samuel R. McKelvie Forest. The idea to grow trees in the treeless Great Plains, conceived around the turn of the century, became a reality in 1902 when the Nebraska National Forest was created.

The Bessey unit of the forest near Halsey contains one of the man-made forests that were planted in the sandhills of central Nebraska. Trees do well here; one scarcely is aware that the forests are not natural. The United States Forest Service considers the Bessey Nursery to be the jewel of the Nebraska National Forest. Since 1902, several million trees have been grown here and transplanted to various national forests across the United States or made available to state governments for distribution. You can drive or walk through the nursery and surrounding plantations and learn more about the organization of the Bessey Nursery. The adjacent Bessey Recreation Area is a modern complex with a campground, swimming pool, and tennis courts—all the comforts of home! A few miles away is the Scott Lookout Tower which provides a view over the extensive man-made forest. The Scott Lookout National Trail can be started here.

The Pine Ridge section of the forest in the northwest corner of Nebraska, south and southwest of Chadron, is the only natural forest and contains native stands of ponderosa pine growing on the slopes and escarpments. Nonetheless, there is grassland here, too; in fact, the ponderosa pines sometimes appear to be growing in a park because of the grassy understory. This forest segment even has a 6,000-acre roadless area west of Fort Robinson, as well as the Topper National Trail. Hikers will appreciate the good trail system here because it reveals a part of Nebraska unknown to many.

Another totally planted unit of forest is known as the Samuel R. McKelvie National Forest. It is sandwiched between the Niobrara and Snake rivers. This area also is sandhill country, with an abundance of native grasses and small ponds. The Steer Campground is comfortable and located at the southwest corner of this region.

SOUTH CENTRAL

There are nine national forests located in the south-central region of the United States. The forests are the Kisatchie in Louisiana; Ouachita, Ozark, and St. Francis in Arkansas; Mark Twain in Missouri; and Angelina, Davy Crockett, Sabine, and Sam Houston in Texas.

Kisatchie National Forest

Location: 598,000 acres in six units in central Louisiana, from below Alexandria to Minden. U.S. Routes 71, 84,

165, and 167 are the major roadways. Maps and brochures available at the district ranger stations in Homer, Pollock, Alexandria, Natchitoches, Leesville, and Winnfield or by writing the Forest Supervisor, 2500 Shreveport Highway, Pineville, LA 71360.

Facilities and Services: Campgrounds (some with camper disposal service); picnic areas with grills; swimming beaches; boat ramps.

Special Attractions: Kisatchie Hills Wilderness; Wild Azalea Trail.

Activities: Camping, picnicking, hiking, boating, swimming, waterskiing, canoeing, hunting, fishing, nature study.

Wildlife: Threatened red-cockaded woodpecker, bald eagle, American alligator.

WHEN YOU LOOK out across the wild Kisatchie Hills from the rocky Longleaf Vista in the Kisatchie National Forest, you may feel the need to check your map to reassure yourself that you are in Louisiana, for this is an area of rugged buttes more characteristic of the wild, woolly West. The hills contain about 8,700 acres of exposed sandstone in west-central Louisiana. Now a designated wilderness area, the Kisatchie Hills offer an unusual opportunity for hiking and viewing scenery. There is an observation shelter at Longleaf Vista that provides a panoramic view of the Kisatchie Hills. You will be surprised to learn that despite the presence of buttes, the maximum elevation above sea level is only 300 feet. Walk from the observation point to a large, flat, white table rock about 100 yards away. This strange formation is surrounded by a band of woody vegetation that includes such shrubs as the colorful fringe tree, which bears white blossoms in April, and the yaupon holly. From the table rock, follow a narrow trail to a conical butte another 100 yards away. Wet seepage areas along the way provide a perfect habitat for an insectivorous sundew and a slender member of the lily family called the colicroot. On the slopes of the butte is the delicate, yellow-flowered smooth indigo. You can wander for several days in this wilderness and probably be surprised often by the variety of vegetation and interesting rocky outcrops.

The other five units of the Kisatchie National Forest are

in more typical Louisiana terrain. There are many low areas that are cypress swamps and wet bayous. On slightly higher ground are pine and hardwood forests. Wild pink azaleas bloom profusely during the latter part of April and have inspired the construction of another popular forest attraction—the 31-mile Wild Azalea National Recreation Trail. Because this special trail is crossed several times by roads, there are many access points for the trail. One 3-mile segment northeast from the junction of Louisiana 489 and Forest Road 273 passes through Boggy Bayou, a wet depression surrounded by dry-land plants like the blackjack oak. Forest Road 287 from Woodworth to Route 488 is particularly scenic where it crosses Castor Creek with its Spanish-moss-draped bald cypresses, glossy-leaved southern magnolias, and saw palmettos.

A major area in the northern unit of the Kisatchie near Minden is the Caney Lakes Recreation Area. Boat ramps are available on both Upper and Lower Caney lakes, while swimming beaches and a waterskiing beach are provided at Lower Caney Lake. The Sugar Cane National Recreation Trail circles Upper Caney Lake. This seven-and-a-half-mile trail goes from bottomland forests with a rich ground covering of wildflowers and ferns to drier pinewoods. Many of Louisiana's 250 kinds of birds may be seen and heard along the trail. Pause at one of the rustic bridges and listen for the calls and chirps of nature.

Fullerton Lake Recreation Area is another nice place to just relax. If you want to, you can walk the Whiskey Chitto Trail through extensive longleaf pine stands. The trail goes through the old abandoned sawmill town of Fullerton where you can still see the remains of several buildings.

When you hike through stands of mature pine, keep alert for the threatened red-cockaded woodpecker. He makes his nest here. At Corney Lake and Saline Lake you might see a bald eagle. A canoe trip can be taken along the Saline Bayou, where in places you float beneath trees laden with wispy Spanish moss.

Ouachita National Forest

Location: 1,575,000 acres in western Arkansas and adjacent eastern Oklahoma, between Hot Springs, Arkansas, and Talihina, Oklahoma; one small unit in southeastern Oklahoma south of Broken Bow, near the Red River. U.S. Highways 71, 59, 27, and 259 are the primary roads penetrating the forest. Maps and brochures available at the district ranger stations in Glenwood, Booneville, Danville, Mena, Oden, Waldron, Perryville, Mount Ida, and Hot Springs National Park, Arkansas, and Heavener, Talihina, and Idabel, Oklahoma or by writing the Forest Supervisor, Federal Building, Hot Springs National Park, AR 71901.

Facilities and Services: Campgrounds (some with camper disposal service); picnic areas with grills; nature center; boat ramps; swimming beaches.

Special Attractions: Talimena Scenic Drive; Robert S. Kerr Memorial Arboretum; several scenic areas; Caney Creek Wilderness.

Activities: Camping, picnicking, biking, boating, canoeing, swimming, waterskiing, hunting, fishing, horseback riding, nature study.

Wildlife: Threatened red-cockaded woodpecker; western animals such as armadillo, roadrunner, scissortail flycatcher.

To GET a preview of the breathtaking, mountainous Ouachita National Forest, take the Talimena Scenic Drive from U.S. Highway 271 north of Talihina, Oklahoma, to Mena, Arkansas. By the time you have completed the 54-mile drive through resplendent and historic mountains and valleys, you will be ready to explore the backcountry of the Ouachita National Forest.

The Talimena Drive is heavily forested along most of the route. Shortleaf pine, black oak, post oak, blackjack oak, and winged elm dominate the drier ridges, while the moister bottomlands contain a rich flora that includes white oak, northern red oak, mockernut hickory, black walnut, basswood, sugar maple, and American beech. If you are fortunate you will glimpse red-tailed hawks, red-shouldered hawks, and turkey vultures soaring overhead. During winter and spring one is apt to spot a golden eagle. Several western animals have made their way into the Ouachita—among those interloping creatures are armadillos, roadrunners, and scissortail flycatchers. In more secluded parts of the forest are bobcats and black bears. There are sweeping vistas along the route, as well as a number of historical sites that include old homesteads, log barns, stone fences, cemeteries, and an old military wagon road.

The Robert S. Kerr Memorial Arboretum, another one of the special attractions, is located about 25 miles east of the western terminus of the drive. (The setting for the novel and subsequent movie *True Grit* is just west of here.) Serving as an outdoor laboratory nature center, the arboretum has three easy trails, each less than 1 mile long. These trails offer a splendid opportunity to see the typical plants that live in the forest. Take time, too, to walk some of the other trails off from the drive. You will see some of the woody plants found here, such as the cucumber magnolia, umbrella magnolia, silverbell tree, bladdernut, and wild hydrangea. If picnic time rolls around while you are on the drive, stop at the Rich Mountain Picnic Area where the panoramic view will give added pleasure to your lunch.

There are several lakes in the Ouachita for those who enjoy any kind of water-based recreation. Two of the most scenic and secluded are 10-acre Lake Silva and 25-acre Shady Lake. Both have swimming beaches and bathhouses.

A float trip on the Ouachita River is becoming increasingly popular each year. The 45-mile route from Pine Ridge (Lum and Abner's town) to Ouachita Lake often passes between high, rocky bluffs. Where the waters are calm, you can fish for smallmouth bass, bream, catfish, and walleye. At other times you will be busy maneuvering through fast-flowing rapids.

There are several undeveloped areas in the Ouachita worthy of a visit. Blowout Mountain near Pencil Bluff, Crystal Mountain northeast of Norman, and Dutch Creek north of Gravelly are designated scenic areas in rugged, mountainous terrain. There are 360-degree vistas from among mature stands of shortleaf pine at each of these areas.

For wilderness and backpacking experience, hike in the 10,000-acre Caney Creek Wilderness in the southern part of the forest. There are unusual shaped rock formations in the Porter, Hanna, Buckeye, and Katy mountains and along clear Caney and Short creeks. Horseback riding is permitted in this wilderness.

We would be remiss if we did not point out the small,

isolated Tiak part of the Ouachita National Forest in southeastern Oklahoma along the Red River near Idabel. Unlike most of the Ouachita which is in the high Ouachita Mountain Province, the Tiak district is in the Gulf Coastal Plain Province. The topography here, by contrast, is generally broad and flat with occasional gently rolling hills; elevation is no more than 530 feet above sea level. The predominant vegetation is more typical of that found in the South—loblolly pine, shortleaf pine, slash pine, willow oak, water oak, cherry-bark oak, and sweet gum. In this district are seven-acre Kulli Lake, which has a swimming beach and picnic area, and the secluded Bokhoma picnic area located near a small pond.

Ozark National Forest

Location: 1,088,000 acres in northwestern Arkansas, mostly between Fayetteville, Fort Smith, and Russellville. U.S. Highway 71 and State Routes 7, 14, 16, 21, and 23 are the major roads. Maps and brochures available at the district ranger stations in Hector, Ozark, Jasper, Paris, Clarksville, and Mountain View or by writing the Forest Supervisor, Box 1008, Russellville, AR 72801.

Facilities and Services: Campgrounds (some with camper disposal service); picnic areas with grills; swimming beaches; boat launching ramps; visitor's information center at Blanchard Springs Caverns.

Special Attractions: Blanchard Springs Caverns; natural bridges; Devil's Canyon.

Activities: Camping, picnicking, hiking, swimming, boating, canoeing, caving, hunting, fishing, horseback riding, nature study.

Wildlife: Endangered or threatened bald eagle and American peregrine falcon.

THE OZARK NATIONAL FOREST is a land of beautiful, rugged mountains. Most of these mountains are in the Boston Mountain Range, but the spectacular Magazine Mountain lies south of the Bostons. High, rocky bluffs, many of them with odd shapes, offer the best panoramic vistas in all of the south-central United States. Hundreds of natural springs give rise to crystal clear mountain streams and spawn uncountable waterfalls. Differential erosion of the rocks has created caves and natural bridges.

The most popular place in the Ozark National Forest is the Blanchard Springs Recreation Area, which features Blanchard Springs Caverns, one of the most unusual caves found in a national forest in this country. Discovered in the 1950s and first opened to public visitation in the late 1960s, the six miles of cave corridors and rooms contain a fantasyland of stalactites, stalagmites, draperies, and other cave formations made from the dripping limestone.

A convenient visitor's information center has been built above the caverns, and it is from here that guided tours are led by forest service personnel at 20- or 30-minute intervals. The caverns are basically in two levels. The tour into the upper level, known as the Dripstone Trail, is shorter and easier than the lower level trail, since it is paved and has ramps at all the inclines. Handicapped persons can be accommodated on this trail. Soft lights have been placed along the way to bring out the rich colors of the formations. The Dripstone Trail is seven-tenths of a mile long.

To reach the lower level Discovery Trail that winds for more than one mile, the visitor must take the elevator from the Visitor Information Center. It descends 216 feet. As you pass various formations, you will have to climb up or down approximately 600 steps to negotiate the 476-foot elevational difference. Down on the lower level you will marvel at the 1,200-by-180-foot Cathedral Room whose ceiling seems to be held up by a 65-foot giant column. A word of warning: Claustrophobic visitors may not enjoy the deeper experience.

After taking the tour through Blanchard Springs Caverns, you may want to linger and hike the trail along North Sylamore Creek, or swim in the creek from the beach near Shelter Cave, or fish in Mirror Lake, or enjoy a campfire program at the rustic outdoor amphitheater. The Blanchard Springs Campground provides a handy base for these activities.

No matter what your interest, visitors are generally impressed by the rock formations in the Ozark National Forest. There are two natural bridges that have been carved out of sandstone. The easily accessible natural bridge at Alum Cove is in a rich hardwood forest. Under one of the sandstone overhangs near the bridge is a colony of rare French's shooting star, whose white or pinkish flowers bloom in late April. Hurricane Creek Natural Bridge is in a less accessible, near-wilderness area, best reached by hiking along a sparkling stream. This 60-foot-high natural bridge is 35 feet long and about 10 feet wide. It is surrounded by a forest of hardwoods, none of which has ever been cut.

One of the wildest and most rugged areas in the Ozark is Devil's Canyon, where the spectacular high cliffs along Mill Creek appear in hues of yellows, reds, and browns. As the sun sweeps overhead, these colors take on different tones. There are piles of exceptionally large rocks along the spring-fed creek, which itself is fascinating because it disappears below ground before reappearing on its way. The canyon is a few miles north of the community of Mulberry.

Several unusual pedestal-based rock formations, caused by the erosive action of rushing water, are scattered throughout the forest. The finest collection of these is along State Route 16 a short distance west of Ben Hur, although Sam's Throne south of Mt. Judea and Three Rocks north of Devil's Canyon are equally fascinating. At Natural Dam a rock barrier placed across Mountain Fork Creek by Mother Nature has formed a pristine, clear mountain pool.

For forest waterfalls, visit Buckeye Hollow, where there are at least seven, or hike along the entire five-mile-long Falling Water Creek which repeatedly cascades over rocky ledges.

St. Francis National Forest

Location: 20,600 acres in eastern Arkansas near the Mississippi River, between Helena and Marianna. State Route 44 goes through the forest. Maps and brochures available at the district ranger station in Marianna or by writing the Forest Supervisor, Box 1008, Russellville, AR 72801.

Facilities and Services: Campgrounds; picnic areas with grills; swimming beaches; boat launching areas.

Special Attractions: Crowley's Ridge, an unusual geological formation.

Activities: Camping, picnicking, hiking, swimming, boating, hunting, fishing, horseback riding.

Wildlife: Threatened bald eagle.

VISITORS will find the St. Francis National Forest small and fascinating. Its fascination lies in the unusual geological formation called Crowley's Ridge. In an otherwise flat terrain, Crowley's Ridge rises conspicuously for a length of 200 miles, from southern Missouri to Helena, Arkansas. The southern tip of the ridge, where the St. Francis National Forest is located, towers nearly 200 feet above the surrounding flatlands. Crowley's Ridge does not have spectacular rock cliffs; instead, it is composed of clay soil, orange sands, and gravel. Much of the heavily eroded slopes have become grotesquely carved. Since this combination of soil, sand, and gravel can be found nowhere in the vicinity, the plants of Crowley's Ridge are also unlike those in the surrounding bottomland forest.

White oak and American beech are the common trees growing throughout the St. Francis. On the shaded, north-facing slopes are handsome tulip poplars, sugar maples, slippery elms, and the rather unusual cucumber magnolia. Here and there is an occasional evergreen American holly, stately basswood with glossy, heart-shaped leaves, butternut with its delicious buttery-flavored fruits, and Kentucky coffee tree, whose seeds at one time were ground and used as a substitute for coffee.

The understory vegetation beneath the forest trees is richer than anyplace else in the region. Small trees and shrubs of ironwood, pawpaw, redbud, flowering dogwood, Hercules'-club, and the sweet-scented spicebush form a middle layer above a dense growth of broad beech fern, Christmas fern, glade fern, golden bellwort, Solomon's seal, and trillium.

There are two recreation areas along the top of Crowley's Ridge, both centered around lakes with swimming beaches and boat launching ramps. The 625-acre Bear Creek Lake, with many secluded necks, is the largest and best developed and also has the best fishing for bass, crappie, and bream. The Beach Point Campground which juts out into Bear Creek Lake is in a restful, parklike setting. Storm Lake, at the southern end of the forest, is a 420-acre lake where fishing is also good.

A small portion of the St. Francis National Forest between Crowley's Ridge and the Mississippi River is flat and is dominated by a dense bottomland hardwood forest with lots of sweet gum, swamp chestnut oak, and other trees that grow in an often saturated soil. If you decide to traipse through these bottomlands, cover yourself well with insect repellent.

Mark Twain National Forest

Location: 1.5 million acres in the southern half of Missouri, primarily between Poplar Bluff and Cassville and from Rolla to the Arkansas border. Interstate 44, U.S. 63, 67, and 180, and State Routes 17, 19, 32, 72, and 125 serve the area. Maps and brochures available at the district ranger stations in Ava, Cassville, Fulton, Doniphan, Fredericktown, Houston, Poplar Bluff, Potosi, Rolla, Salem, Van Buren, Willow Springs, and Winona or by writing the Forest Supervisor, 401 Fairgrounds Road, Rolla, MO 65401.

Facilities and Services: Campgrounds (some with camper disposal service); picnic sites with grills; boat and canoe access areas; off-road vehicle areas.

Special Attractions: Unusual Ozark plants; clear gravel-bottomed streams; limestone glades; several designated wilderness areas.

Activities: Camping, picnicking, hiking, hunting, fishing, swimming, boating, canoeing, floating, horseback riding, off-road vehicle riding, nature study.

Wildlife: Threatened bald eagle.

MY FAMILY in my childhood years would often pack a picnic basket and head across the Mississippi River into the Missouri Ozarks. These were special days, partly because the Ozarks were the only mountains this writer had ever seen, and partly because the crystal clear, gravel-bottomed streams were refreshingly unlike the muddy creeks in southern Illinois where this author grew up. Sometimes we would explore around the old abandoned silver mines; on longer jaunts we would wander up and down the scenic Eleven Point River, with perhaps a short side trip to Greer Springs. It was not until several years later, after embarking upon a scientific career, that this writer realized the Ozarks were an outdoor laboratory of remarkable plants, animals, geology, and unparalleled scenery.

Much of the flavor of the Missouri Ozarks has been preserved in the Mark Twain National Forest, a sprawling forest composed of nine noncontiguous units across the southern half of Missouri and extending all the way to the Arkansas border. The Ozarks are one of the oldest mountain ranges in the country. Their worn-down, rounded peaks are densely clothed by a vegetation dominated by shortleaf pine and upland oak species. Delicate pink azaleas bloom beneath the yellow-green pines during mid-May, providing an eye-catching composition to the visitor. Despite the presence of secluded dwellings by homesteaders, the general impression of the Ozarks is one of an unbroken forest interrupted occasionally by granite or limestone outcrops. Winding quietly in the valleys are the clear streams, where fishing, canoeing, and floating are popular. Favorite among these are the Current, Eleven Point, and Black rivers. Some of the rivers have been dammed, furnishing recreation areas such as the popular Table Rock Lake and Lake Taneycomo.

Several features of the Mark Twain make it unique among national forests. For one thing, piles of huge boulders occasionally jam up along the course of a creek to form what the local residents call "shut-ins." Not only are these shut-ins exceptionally scenic, but they also have rather rare plant species, such as the shining blue-star, associated with them.

Even more botanically significant are the forest's limestone glades. These are essentially treeless open areas usually found on the upper slopes of the ridges where limestone rocks have come to the surface and are surrounded by low-growing vegetation. The showiest of the glade plants is the Missouri evening primrose, an herb that forms bright yellow flowers up to six inches across during May. Another gem on some of the limestone glades is the giant purple foxglove, with two-and-a-half-inch-long flowers that also usually bloom in May. Then there is the thick, leathery, bell-shaped Ozark clematis, another early spring bloomer restricted to the Ozarks. All in all, botanists have recorded dozens of plants, many of them found nowhere else.

The most spectacular of these glades is Hercules Glade. Anytime after frost is a good time to visit, but our favorite times are early May, when the evening primrose and fox-glove flower, or late summer, when the coneflowers and the rosy pink palafoxia bloom. To get to Hercules Glade, leave from Ava, taking the primitive but highly scenic forest road known as the Gladetop Trail.

You will surely want to hike along or canoe in some of the Ozark rivers. In February look in the gravel along the rivers for the yellow blossoms of the vernal witch hazel, a tree endemic to the Ozarks and which flowers when every-thing else is dormant.

For a never-to-be-forgotten experience, we suggest a canoe or float trip along the Eleven Point River, one of the nation's designated Wild and Scenic rivers. The 44-mile stretch from Thomasville to the Narrows at Missouri Route 142 will take more than a day, but there are several good forest service campgrounds along the way. You can float, of course, for shorter distances to coincide with the time available. The river sometimes flows between 300-foot sheer, vertical cliffs. Among the spectacular cliffs are those at Cane Bluff and the Narrows. The river also passes by more than 30 springs that feed it. Look for misty Blue Spring near the Narrows. A word of caution: There are a few fast water chutes to be aware of, such as the Mary Decker Shoal and the Halls Bay Chute.

Visitors to the Mark Twain are also offered the oppor-tunity to see at the Silver Mines Recreation Area along the St. Francis River the remnants of the Einstein Silver Mine that was abandoned in 1946. There remain prospect pits, dikes, ore veins, concrete bulwarks, slag piles, and even a mine opening with a wooden alcove. A trail permits you to get a closer look at these sites, as well as pass one of the shut-ins of the St. Francis River. There are a number of enjoyable campsites in the area. A few miles southeast of Silver Mines is Marble Creek, another developed camping and picnicking site. Still evident along the creek is the "marble," in reality colored dolomite that used to be mined and sold as marble. You can also see remains of an old grist mill dam.

Sutton's Bluff, a huge, isolated, sheer-faced cliff that

suddenly rises spectacularly along the clear-flowing West Fork of the Black River in Peaceful Valley, is a landmark you will want to see in the Mark Twain.

For a pleasant hiking experience, take the Ridge Runner Trail. Although the entire trail is 23 miles long, there are several access points that permit shorter segments to be hiked. You get close to the Ozarks along this route. Hell-roaring Spring near the Noblett terminus of the trail is a sterling example of a pure Ozark spring. At Blue Hole, where every morning dawns in a hazy mist, there is a haunting feeling of mysterious beauty. Stream Mill Hollow is one of several dark, deep hollows along the trail that is penetrated by a network of rocky-bottomed spring branches. The two ends of the trail are a few miles west of Willow Springs and West Plains, respectively.

Four wilderness areas ranging in size from the 3,920-acre Rock Pile Mountain Wilderness to the 12,315-acre Hercules Glade Wilderness provide ample opportunity for backcountry exploration. The summit of Rock Pile Mountain, with an unexplained pile of granite rocks arranged in a circle, is reached only after hiking through a never-cut forest of basswood, Kentucky coffee tree, butternut, black walnut, and several kinds of oak.

Although Bell Mountain dominates the 8,530-acre wilderness that bears its name, the most interesting feature in this back area is Shut-in Creek, where several examples of remarkable shut-ins occur in rapid succession. Devil's Backbone Wilderness, a dry sharp ridge of 6,800 acres that rises steeply above the North Fork of the White River, is noted for its brilliant display of wild pink azaleas in early May.

In addition, there are miles of horseback riding and authorized off-road vehicle trails in the Mark Twain. The Big Piney Riding Trail is a 17-mile trail from the Big Piney Trail Camp on Highway 17 to the community of Roby. One developed area for off-road vehicles is the Chadwick Motorcycle Trail Area, complete with campgrounds, and located southeast of the town of Ozark.

Angelina National Forest

Location: 155,000 acres in east-central Texas between Hemphill and Lufkin. U.S. Highway 69 and State Routes

83, 47, 103, and 63 cross the forest. Maps and brochures available at the district ranger station in Lufkin or by writing the Forest Supervisor, Federal Building, Box 969, Lufkin, TX 75901.

Facilities and Services: Campgrounds (some with camper disposal service); picnic areas with grills; swimming beaches; boat launching areas.

Special Attraction: Sam Rayburn Reservoir.

Activities: Camping, picnicking, hiking, swimming, boating, waterskiing, hunting, fishing.

Wildlife: Threatened red-cockaded woodpecker.

THE SMALL ANGELINA NATIONAL FOREST, located midway between the Sabine and Davy Crockett national forests, surrounds the 114,000-acre Sam Rayburn Reservoir. The reservoir is one of the best in the country for bass.

Our favorite area in the forest is the Caney Creek recreation site on the Sam Rayburn Reservoir. Although you can swim, boat, water-ski, or fish here, we were intrigued during our recent visit by the abundance of early spring wildflowers. Blue salvias were blooming everywhere; occasional patches of the yellow lousewort, the pale blue bird's-foot violet with its deeply cleft leaves that resemble a bird's foot, and the beautiful magenta-flowered poppy mallow provided additional color. An unusual interpretive trail called the Board Walk is not what one might think. Instead of a trail that crosses boggy areas by means of a boardwalk, the trail has exhibits along it explaining about different kinds of boards—their measurements, the way they are cut, etc.

Dry areas with deep, sandy soil within the forest have a parklike appearance where longleaf pine is the dominant tree, although shortleaf and loblolly pines grow here occasionally, as do bluejack oak, blackjack oak, and sand post oak. Where wet areas occur among the longleaf pine stands, plants such as red maple, sweet gum, red bay, and sweet bay may be found.

On the dry, wooded ridges, the vegetation is an attractive mixture of loblolly pine, shortleaf pine, post oak, white oak, black oak, blackjack oak, sour gum, and a number of hickories. In the heavily shaded, wet areas adjacent to streams are dense growths of loblolly pine, sweet gum, water oak, willow oak, swamp chestnut oak, cherry-bark oak, overcup oak, and the American beech. In the wettest sites are bald cypress, tupelo gum, water hickory, red maple, and sugarberry. In some of the older growth pine forests, the threatened red-cockaded woodpecker is found. Unfortunately we did not see any the week we were there.

If there is time, stop at Bouton Lake, Letney, and Boykin Springs, all areas with campgrounds and picnic tables. The 7-acre Bouton Lake is in an old river channel surrounded by a fine forest of hardwoods. Letney, by contrast, is more of an upland site with an abundance of pines. From this area near the south end of the Sam Rayburn Reservoir are excellent vistas of the lake. Boykin Springs is a 10-acre, spring-fed impoundment in a setting of longleaf pine.

Davy Crockett National Forest

Location: 161,000 acres in east Texas, a few miles west of Lufkin. U.S. Highway 287 and State Routes 7, 21, and 94 serve the forest. Maps and brochures available at the district ranger stations in Crockett and Apple Springs or by writing the Forest Supervisor, Federal Building, Box 969, Lufkin, TX 75901.

Facilities and Services: Campgrounds (camper disposal service at Ratcliff Lake); picnic areas with grills; swimming beach; concessions.

Special Attraction: Four C National Recreation Trail.

Activities: Camping, picnicking, hiking, swimming, boating, canoeing, hunting, fishing.

Wildlife: Threatened red-cockaded woodpecker.

A SUBTLE CHANGE in the forest takes place as you leave the Sabine and Angelina national forests in eastern Texas and head west into the Davy Crockett. There is a gradual decrease in the number of pines and an increasing amount of oaks, particularly the tough, leathery-leaved post oak.

The twistingly scenic Neches River forms the eastern boundary of the Davy Crockett National Forest. Neches Bluff at the north end of the forest near State Route 21 has a commanding view of the Neches River bottomlands below and the pine and hardwood forest on the bluffs. Here is a good place to picnic, also, although when we were there, a chattering squirrel in a tree overhead kept pelting us with nut fragments.

Neches Bluff is also one end of the 20-mile Four C National Recreation Trail. We heartily recommend this trail if you have time and stamina, because it provides an excellent cross section of the Davy Crockett. Dropping away from the high bluff above the river, the trail alternates between pine-forested ridges and bottomland hardwoods, and upland forests and boggy sloughs. A number of footbridges are provided to keep you out of the wettest areas. The primitive Walnut Creek and Pond campsites, which have no modern conveniences, are spaced along the trail.

The Four C Trail's southern terminus is at the Ratcliff Recreation Area, the most developed area in the Davy Crockett. A former log pond of the Central Coal and Coke Company sawmill is now a 45-acre lake suitable for swimming, boating, and fishing. There is a concessionaire near the swimming beach. Nearby is the old historic town of Ratcliff, where a few businesses still flourish.

A meandering and fun canoe trail has been developed that utilizes the Neches River and a swampy slough. After putting in at the canoe trailhead at Scurlock Camp, canoe in a clockwise direction to take advantage of the direction of the current. You will soon enter the Neches River and float southward until you cut inland to the Big Slough. You suddenly experience the solitude of being in a much larger wilderness.

Although the Ratcliff Recreation Area has all the modern amenities for camping and recreation, you can find peace at the small, secluded Holly Bluff Campground overlooking the Neches River. From here you can watch the sun rise above the river in the yellow morning haze.

Sabine National Forest

Location: 187,000 acres in east-central Texas on the Texas-Louisiana border near Hemphill and San Augustine. U.S. Highway 96 passes by the western edge of the forest. Maps and brochures available at the district ranger stations in San

Augustine and Hemphill or by writing the Forest Supervisor, Federal Building, Box 969, Lufkin, TX 75901.

Facilities and Services: Campgrounds (some with camper disposal service); picnic areas with grills; boat ramps; swimming beaches.

Special Attraction: Toledo Bend Reservoir.

Activities: Camping, picnicking, hiking, swimming, fishing, boating, waterskiing, hunting.

Wildlife: Threatened red-cockaded woodpecker and American alligator.

MOST OF the recreation activities in the Sabine National Forest center around the 181,000-acre Toledo Bend Reservoir, formed by the damming of the Sabine River. The Sabine National Forest maintains 192 miles of shoreline along the reservoir.

We like the Red Hills Lake area as a base for activities in this forest. Red Hills Lake is a 17-acre lake fed by natural springs. There is a swimming beach and bathhouse, and boats without motors are permitted on the lake. Fishing for bluegill and crappie is good. The Dogwood Campground near the lake is in a pleasant woodland setting. There is ultimate beauty in late April when the flowering dogwoods and redbuds bloom. Even when the dogwoods are not flowering, the loblolly pines, sweet gums, and smooth-trunked American beeches emit a freshness when you awake in the morning.

You can hike from this area to Chambers Hill, gradually climbing into a drier upland habitat where the vegetation abruptly changes to a hardwood forest with Spanish oaks, winged elms, and sour gums present. Small, shrubby, dazzling red buckeyes, some blooming when only a foot tall, add color along the trail during mid-April. Although the trail is pleasant, there is an abundance of poison ivy in the lower areas. Insect repellent is a must because mosquitoes are usually present.

Other major recreation areas at opposite ends of the forest are Ragtown and Willow Oak. Both areas have campgrounds, boat ramps, and an interpretive hiking trail. The mile-long trail at Ragtown is more than twice as long as the trail at Willow Oak.

A more restful area can be found along Patroon Creek north of Geneva. Fishing in the nearby stream, or just sitting under the trees, can provide for uninterrupted relaxation. East of Patroon Creek is a wild area known as Chambers Ferry. If you hike this area to the Toledo Reservoir, you will come to 50-foot bluffs. From these bluff tops are breathtaking views of the reservoir and the surrounding countryside.

Another unusual region in the Sabine National Forest lies along Colorow Creek, about three miles north of Geneva. Here is a natural sandstone bridge, as well as deep, rocky stream channels.

Finally, make your way down to Indian Mounds in the southeastern corner of the forest and ponder over the small, mysterious mounds in the region, remnants from an earlier culture. A campground is nearby.

Throughout the Sabine there is a chance you might encounter the threatened red-cockaded woodpecker. There are also American alligators at Housen Bayou, Hurricane Bayou, and between Indian Mounds and Harpers Ridge.

Sam Houston National Forest

Location: 158,000 acres in southeast Texas, near Huntsville and Cleveland. Interstate 45 bisects the forest. Maps and brochures available at the district ranger stations in Cleveland and New Waverly or by writing the Forest Supervisor, Federal Building, Box 969, Lufkin, TX 75901.

Facilities and Services: Campgrounds (some with camper disposal service); picnic areas with grills; boat ramps; swimming beaches.

Special Attraction: Lone Star National Recreation Trail.

Activities: Camping, picnicking, hiking, backpacking, hunting, fishing, swimming, boating.

Wildlife: Threatened red-cockaded woodpecker.

THE BEST WAY to experience all of the Sam Houston National Forest is to hike the Lone Star National Recreation Trail, a 104-mile route that crosses the forest from Longstreet on the western boundary nearly to the town of Cleveland in the southeast corner of the forest. You can, of course, join the trail anywhere, and there are several optional loops you can take to prolong your hike. Since most of the trail is through flat terrain, there are no climbing problems, but there are a few hazards to beware of. Hiking during the hot and humid summer is not recommended, and you should be prepared in spring and autumn to encounter poison ivy, poisonous snakes, ticks, mosquitoes, and chiggers. There are also places where you may have to wade across creeks, or find a fallen log to get across. The charm of the area is the lush, dense vegetation that crowds in all around you. One of the best-developed sites along the trail is Stubblefield on the shore of Lake Conroe. There is a good campground here, and a short nature trail has been laid out that lets you become familiar with some of the plants of the area.

Another well-developed area is at Double Lake where you can cool off by swimming. At your campsite in the evening, breezes usually waft gently through the treetops. The Lone Star Trail south from Double Lake passes through the Big Creek Scenic Area, a region of lush magnolias, pines, oaks, and a wide assortment of other trees, shrubs, and herbs. Near the trail's end is Winter Bayou, a wet area filled with moisture-loving plants unlike those seen elsewhere along the trail.

NORTH ROCKIES

The largest number of national forests in all of the United States is located in the sprawling north Rockies. There are a total of 27 forests. In this region are the Beaverhead, Bitterroot, Custer, Deerlodge, Flathead, Gallatin, Helena, Kootenai, Lewis and Clark, and Lolo in Montana; Black Hills in South Dakota; Big Horn, Bridger-Teton, Medicine Bow, and Shoshone in Wyoming; and Boise, Caribou, Challis, Clearwater, Coeur d'Alene, Kaniksu, Nezperce, Payette, Salmon, Sawtooth, St. Joe, and Targhee in Idaho.

Beaverhead National Forest

Location: 2,114,577 acres in southwestern Montana, southwest of Butte. Interstate 15, U.S. Highway 93, and

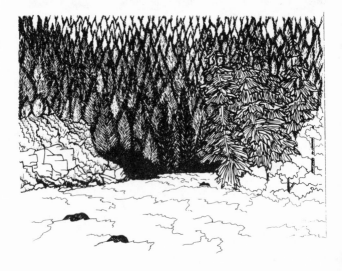

State Routes 43, 278, and 324 serve the forest. Maps and brochures available at the district ranger stations in Dillon, Ennis, Sheridan, Wisdom, and Wise River or by writing the Forest Supervisor, 610 N. Montana St., Dillon, MT 59725.

Facilities and Services: Campgrounds (some with camper disposal service); picnic areas with grills; boat ramps.

Special Attraction: Anaconda-Pintlar Wilderness.

Activities: Camping, picnicking, hiking, backpacking, hunting, fishing, boating, winter sports, horseback riding, nature study.

Wildlife: Endangered or threatened bald eagle, American peregrine falcon, Rocky Mountain wolf, and grizzly bear; wolverine, mountain lion, bobcat, mountain goat, Rocky Mountain bighorn sheep, moose, elk, golden eagle.

IF YOU ARE A HIKER who enjoys reliving pioneer days, you can travel over the same route that Lewis and Clark did more than 175 years ago. In the Beaverhead Forest you will find Beaverhead Rock, the landmark used by Sacajawea, Lewis and Clark's Indian guide. Lemhi Pass, the place where Lewis first crossed the Continental Divide, is also in the forest and can be reached by back road or hiking trail. You will need a map to get there. The small spring at the pass was described by Lewis as the "headwaters of the mighty Missouri." The pass is beautiful during the summer when a myriad of colorful wildflowers—lupines, wyethias, paintbrushes, cinquefoils—are in full bloom. As you look westward from it, as Lewis himself must have done when he was there on August 12, 1805, it is an awesome sight. You can see towering mountain after towering mountain all the way to the horizon.

Much has changed in the 175 years that have passed since the Lewis and Clark journey. Gold was discovered, smelters and charcoal kilns were built, dusty mining towns and then ragtag cities sprang up, Indians were chased out, and highways and other infrastructure came about.

What has not changed is the natural setting found in the Anaconda-Pintlar Wilderness, the forest's key attraction. About half of this pristine 158,516-acre wilderness is in the

Beaverhead, the remainder being divided among the Bitterroot and Deerlodge national forests. Rugged peaks, alpine lakes, basins carved by glaciers, and broad valleys are characteristic features of this unspoiled area. Many trails offer you the opportunity to penetrate the wilderness. The one from the Pintlar Campground to Pintlar Pass on the Continental Divide passes through a variety of plant habitats. The campground, situated along Pintlar Creek and near Pintlar Lake, is an idyllic place to camp, with the incessant but soothing sound of its rushing creek to lull you to sleep. Shortly after entering the wilderness from the campground, you will hear Pintlar Falls and then encounter its bubbling white waters. You next will come upon Pintlar Meadows; high peaks of the Continental Divide form a majestic backdrop. After a steady uphill climb east of the two Pintlar Peaks, you will reach Pintlar Pass. There is a chance that elk may be grazing in the meadows or mountain goats on the high peaks. As for plant life, in addition to Douglas fir, Engelmann spruce, and subalpine fir, you can see the less common subalpine larch and whitebark pine. In this section of the Anaconda-Pintlar is also West Goat Peak, at 10,793 feet, the highest elevation in the wilderness.

Over the years we have seen many beehive-shaped charcoal kilns on our travels. Near Canyon Creek in the forest is an opportunity to see some fine examples. These kilns, built for the copper smelter at Glendale about 1880, are nestled at the base of a steep, conifer-dotted slope and are reached via road and trail a few miles west of Interstate 15.

East of Interstate 15 are the Gravelly, Tobacco Root, and Madison ranges. There is a scenic loop road in the Gravelly Range that encircles 10,547-foot Black Butte. A guide brochure to this 60-mile loop is available at the ranger stations. The Madison Range includes Hilgard Peak, the highest mountain in the forest, as well as some amusingly shaped peaks that have earned them the names of Sphinx Mountain and The Helmet.

There is an abundance of wildlife in the forest, including several large mammals, among which are the endangered grizzly bear and Rocky Mountain wolf. We saw neither, but we were lucky enough to see our first pygmy rabbit, a pint-sized hare that is more common south of the forest but rare this far north.

Bitterroot National Forest

Location: 1,577,930 acres in west-central Montana and east-central Idaho, south of Missoula and on either side of Hamilton. U.S. Route 93 bisects the forest from north to south. Maps and brochures available at the district ranger stations in Darby, Stevensville, and Sula, or by writing the Forest Supervisor, 316 N. 3rd St., Hamilton, MT 59840.

Facilities and Services: Campgrounds (some with camper disposal service); picnic areas with grills; boat ramps; swimming beaches.

Special Attractions: Selway-Bitterroot Wilderness; Anaconda-Pintlar Wilderness; River of No Return Wilderness.

Activities: Camping, picnicking, hiking, backpacking, hunting, fishing, boating, swimming, rafting, horseback riding, nature study, winter sports.

Wildlife: Black bear, elk, moose, bobcat, mountain lion, Rocky Mountain bighorn sheep, mountain goat.

THREE WILDERNESS AREAS are musts for the visitor to the Bitterroot. The first is the Selway-Bitterroot Wilderness whose focal point is the Bitterroot Range, a group of high jagged peaks to the west of the wilderness. The broader topped mountains eastward include portions of the Anaconda-Pintlar Wilderness. Summer visitors are attracted to the Bitterroots and their lofty aeries. Among the popular peaks are 10,157-foot Trapper, slightly lower North Trapper, El Capitan, Sugarloaf, Como, and The Lonesome Bachelor. To get an eagle's view of this spectacular range, we drove the eight-mile, rocky, one-lane gravel road that leaves U.S. 93 between Hamilton and Darby to the road's end at the Lost Horse Observation Point. Amid cliffs and over nests of alpine wildflowers you can look across to El Capitan and the Como Peaks, see a classic U-shaped valley, and catch a glimpse of some churning water deep in the canyon below.

The second wilderness area of note is the Anaconda-Pintlar Wilderness. However, only a small segment of this untamed land is here; the remainder sprawls over parts of the Beaverhead and Deerlodge national forests. Many high lakes dot the western slopes of the mountains in the Bitterroot part of the wilderness. Kelly, Ripple, and Hidden lakes, all near the Bitterroot Pass, are among the more scenic and can be reached by good hiking trails. Cutthroat and rainbow trout are awaiting anglers and their frying pans. If you hike the Bitterroot River Trail from the East Fork Guard Station to Hidden Lake, you will pass roaring Star Creek Falls near the junction of Star Creek and the Bitterroot River.

The third wilderness on the forest list of sights is the River of No Return, fully treated in the Nezperce Forest entry. The segment in the Bitterroot contains a scenic but rugged trail that connects Eakin Point with impressive 8,944-foot Salmon Mountain.

If you are prepared for back road adventure, there are some further thrilling experiences to savor. Take the never-to-be-forgotten road (Route 473) that leaves U.S. 93 4 miles south of Darby and tortuously winds for many miles to the Paradise Campground on the wild and ferocious Selway River. There are breathtaking vistas at every turn. On this route you also get the flavor of the Bitterroot and will find it a good jumping-off place for other forays. The first 15 miles are paved as the road keeps pace with the West Fork of the Bitterroot River. Just below the ranger station, you must take the gravel road (still 473) that leads west from the main road. (The paved road continues south to Little Boulder Bay where there is swimming and boating, and farther on to a special area for handicapped persons at Alta Pine National Recreation Trail.) The gravel road follows the Nezperce Fork past Peyton Rock and the Fales Flat Group Campground before it twists its way over Nezperce Pass. After descending the western slope, the road begins a curving route that stays near clear-flowing Deep Creek. As the road comes around to the north of Roundtop Mountain, a branch coils its way south through spectacularly scenic country over Hells Half Acre Saddle and Vance Creek Saddle to Hells Half Acre Mountain. If you are so inclined, there are several trails you can take off this branch road, including the Storm Ridge Trail to Devils Point. Returning to the gravel route, continue west as the road intersects the Selway River and begins a course northward until it crosses the river at Magruder Crossing. One mile west by trail is the site of the bloody 19th century Magruder Massacre. From Nezperce Pass to Magruder Crossing, the road follows a narrow course that separates the Selway-Bitterroot to the north from the River of No Return Wilderness to the south. North from Magruder Crossing to the Paradise Campground, the road continues along the Selway River and forms a narrow protrusion into the Selway-Bitterroot Wilderness. If you have time, take the south fork from Magruder Crossing. The road winds over Haystack Saddle, Kim Creek Saddle, and Magruder Saddle to a breathtaking observation point across from Deadtop Mountain. After viewing the wild western terrain, continue on past Salmon Mountain to the Sabe Vista southwest of Sabe mountain, where an overview is provided of the beauty of the Selway-Bitterroot Wilderness to the north and the River of No Return Wilderness to the south.

There is also a four-mile historical trail in the forest that leads from U.S. Route 93 six miles south of the Sula Ranger Station to Gibbons Pass. It follows the route used in 1877 by Chief Joseph and the Nezperce tribe as they sought an escape route to Canada. A few blazed trees, said to be the work of Indians, can be seen along the trail.

Custer National Forest

Location: 1,186,000 acres in southern Montana and northwestern South Dakota, south and east of Billings, Montana. U.S. Highway 212 crosses much of the forest. Maps and brochures available at the district ranger stations in Ashland and Red Lodge, Montana, and Camp Crook, South Dakota, or by writing the Forest Supervisor, Box 2556, Billings, MT 59103.

Facilities and Services: Campgrounds (some with camper disposal service); picnic areas with grills; boat ramps.

Special Attraction: Absaroka-Beartooth Wilderness.

Activities: Camping, picnicking, hiking, backpacking, hunting, fishing, horseback riding, boating, nature study.

Wildlife: Rocky Mountain bighorn sheep, mountain goat, elk, moose, black bear, mountain lion.

THE CUSTER NATIONAL FOREST is divided into several distinct units across southern Montana and northwestern South Dakota. At the western extreme are the Beartooth Mountains, an area of high rugged peaks and deep canyons. Eastward, the mountains are lower and eventually become almost gentle hills.

The Beartooth Mountains are largely in the Absaroka-Beartooth Wilderness—the forest's major visitor attraction. Of unusual interest is Grasshopper Glacier at the edge of the wilderness 8 miles north of Cooke City. The glacier, which clings to the side of a mountain at 11,000 feet, is about 1 mile long and ½ mile wide. It is significant in that buried in the ice are millions of grasshoppers that biologists believe swarmed over the mountain range two centuries ago. It is speculated that they were caught in a severe storm and somehow landed on the glacier where they were immediately covered by biting snow and ice. From Grasshopper Glacier you are only 3 miles to Granite Peak, Montana's highest mountain at 12,799 feet. It is a rugged ascent, and only experienced climbers should attempt to conquer this peak. For the more conventional hiker, one of the most rewarding trails in the wilderness leaves the Spread Creek Campground at Rosebud Lake and winds and climbs for nearly 6 miles to Rainbow Lake. Nestled at the base of sharp cliffs and surrounded by a forest of fragrant pines, the lake is a good place to relax and take in the surrounding beauty and sights of the region. Outside the wilderness, we recommend a trip to the Pryor Mountains about 20 miles south of Billings. A highlight is seeing deep Crooked Creek Canyon, whose perpendicular cliffs stand 2,200 feet above the narrow chasm. Nearby are several ice caves worth visiting, where temperatures remain so cold that ice persists throughout the year.

Another side trip that is entertaining is one to Capitol Rock. At the extreme eastern edge of Montana, in the Long Hills unit of the forest, you will find this formation, which has a striking resemblance to the nation's capitol in Washington, D.C., when viewed from a distance. The Wickham Campground nearby is a good base from which to explore this section.

The South Dakota portion of the Custer contains several strange geological formations of note. One of them is the

Honey Combs northeast of the Long Hills. It is a series of gigantic vertical folds that rise to varying heights and resemble the hexagonal cells that bees build to store their honey. Another formation of interest is at nearby Reva Gap. Here you will see the so-called Slim Buttes, several lean and dry crags that rise out of an otherwise flat terrain. You can wander for hours among other strange formations in the area. We conjured up all sorts of fanciful names for them.

Deerlodge National Forest

Location: 1,195,754 acres in western Montana, west, north, and east of Butte. Interstates 15 and 90, U.S. Highway 10A, and State Routes 38, 55, and 69 are the access roads. Maps and brochures available at the district ranger stations in Butte, Deer Lodge, Whitehall, and Philipsburg or by writing the Forest Supervisor, Federal Building, Box 400, Butte, MT 59703.

Facilities and Services: Campgrounds (some with camper disposal service); picnic areas with grills; boat launching areas; swimming beaches.

Special Attraction: Anaconda-Pintlar Wilderness.

Activities: Camping, picnicking, hiking, backpacking, hunting, fishing, boating, swimming, horseback riding, nature study.

Wildlife: Black bear, moose, elk, mountain lion, bobcat, Rocky Mountain bighorn sheep, mountain goat.

SEVERAL ISOLATED MOUNTAIN RANGES make up the Deerlodge National Forest. The impressive Anaconda Mountains form the core of the Anaconda-Pintlar Wilderness, a special attraction of the forest, which is partly in the Deerlodge, Bitterroot, and Beaverhead national forests. A 45-mile Highline Trail follows the Continental Divide through pristine alpine terrain in this wilderness, passing at least 10 peaks in the 9,500-to-10,400-foot range. Warren Peak, with nearby Tamarack Lake, offers a particularly handsome panorama in the Deerlodge portion of the wilderness.

The Sapphire Mountains make up a beautiful range due west of Anaconda that forms the western boundary of the

Deerlodge. There are several back roads in the Sapphire Range that permit easy access to the mountains. One good way to cross the Sapphires is to leave the West Fork Guard Station and eventually climb to Shalkaho Pass. If you wish to stay overnight, camp at the small Crystal Creek Campground on the approach to the pass.

The Highland Mountains start their ascent a few miles south of Butte and continue to climb until they soar above 10,000 feet at Red Mountain and Table Mountain. A road from Pipestone Pass on U.S. Highway 10A winds through the Highland Mountains to the ghost town of Highland City. From the ruins and the old cemetery there are trails south to Red Mountain and west to Mt. Humbug.

In the Elkhorn Mountains the ghost town of Elkhorn City was on our itinerary. We proceeded on a 12-mile gravel road from State Route 281 southeast of Boulder. In a short while we crossed the boundary into the forest and made our way through scenic parklike openings and coniferous forests toward Elkhorn City. Suddenly the road came to a clearing. Ahead on the hill were the weathered, dark brown frame buildings that remain of this once booming Elkhorn City. Most of the buildings are now windowless, and a bullet hole or two can be found in some of the walls, but the Grand Hotel, Fraternity Hall, and several lesser buildings still stand, thanks to the efforts of the Montana Historical Society. There are still residents in this ghost town, so take care not to trespass on private property. It is, of course, illegal to molest the antiquated structures that still stand.

To visit the Tobacco Root Range we retraced our steps to Route 281 and proceeded south toward Interstate 90. When we rounded a sweeping curve, the front range came into view, still snowcapped in late June. As we started closing the 30-mile gap to the mountains, we could identify several of the 10,000-foot peaks—Mount Jefferson, Horse Mountain, Hollow Top Mountain, Middle Mountain, Grants Peak, Noble Peak, Mt. Jackson, and Lakeshore Mountain. These lofty landmarks form the border between the Deerlodge and Beaverhead forests. There is a rough road—if you feel up to it—that penetrates into this range, ending near Sailor Lake at the northern foot of Mt. Jackson and Lakeshore Mountain. By comparison, the Flint Creek Mountains are generally lower and more round topped than the other forest mountains, although rugged Mt. Powell reaches an elevation of 10,164 feet. There is an interesting hiking trail in the range that leaves from the south end of Rock Creek Lake (not in the forest), veers to the east of East Goat mountain, and zigzags its way over Racetrack Pass, providing access to a string of clear mountain lakes — Elbow Lake, the Trask Lakes, Alpine Lake, Albicaulis Lake, and Dead Lake. Bubbling Rock Creek Falls is a short distance from the trail beginning.

Flathead National Forest

Location: 2,350,000 acres in northwestern Montana. U.S. Highways 2 and 93 and State Routes 35, 82, and 83 are the principal roads. Maps and brochures available at the district ranger stations in Columbia Falls, Hungry Horse, Big Fork, and Whitefish, at the Hungry Horse Reservoir Visitor's Center, or by writing the Forest Supervisor, 1935 3rd Ave. E, Kalispell, MT 59901.

Facilities and Services: Campgrounds (some with camper disposal service); picnic areas with grills; boat ramps; swimming beaches; visitor's center.

Special Attractions: Bob Marshall Wilderness; Great Bear Wilderness; Mission Mountains Wilderness; Jewel Basin Hiking Area.

Activities: Camping, picnicking, hiking, backpacking, hunting, fishing, boating, swimming, horseback riding, nature study, winter sports.

Wildlife: Endangered or threatened grizzly bear, bald eagle, and American peregrine falcon; black bear, Rocky Mountain bighorn sheep, mountain goat, wolverine, mountain lion, bobcat, elk, moose.

THE VAST WILDERNESS of the Flathead National Forest is a paradise for hikers and backpackers. The western side of its prime 950,000-acre Bob Marshall Wilderness, which we consider one of the finest wilderness areas in the country, is here, with its spectacular high walls, mountain lakes, and waterfalls. There is also Big Salmon Lake, which is about

5 miles in length. A good trail to it leaves from Holland Lake and follows Smoky Creek and Big Salmon Creek to the lake, passing en route Big Salmon Falls. On your return we suggest you follow the South Fork Flathead River to the Big Prairie Ranger Station (in the wilderness), then continue east along Bartlett Creek, Show Creek, and Lick Creek before crossing Gordon Pass and dropping down to Holland Lake. The forest service estimates this loop to be 70 miles and to take a minimum of seven days. One side trip worth taking is down Una Creek 8 miles to Bullet Nose Mountain where there are large deposits of fossil trilobites, small animals from an earlier era.

If you are not up to long hikes, there are two shorter ones we suggest. One is from the Silvertip Guard Station about seven miles into the narrow Spotted Bear River gorge where wispy little Dean Falls drops into a trout-filled basin. Be sure and take your fishing rod with you. The other, somewhat longer hike follows rough and rocky Gorge Creek to oblong-shaped Sunburst Lake, a glacier-fed lake formed from a huge active glacier on the north flank of Swan Peak. The complete panorama of mountain peak, glacier, and lake is hard to surpass.

Great Bear Wilderness, another forest must, lies across U.S. Highway 2 from Glacier National Park. It is composed of 286,000 acres of glacier-carved terrain with U-shaped valleys, 8,000-foot mountains, above-timberline vegetation, and glaciers. Both Great Northern Mountain and Mt. Grant in this area have glaciers on their eastern face; they are accessible by eight- and six-mile trails, respectively, from U.S. Highway 2.

Mission Mountains Wilderness, a further forest feature, is a narrow strip of wild country a few miles from the southeast corner of Flathead Lake. This 74,000-acre wilderness occupies the eastern slope of the Mission Mountains. Since there are few developed trails, the area should be reserved only for the most experienced hikers who do not mind going cross-country.

Our favorite area in the Flathead is the Jewel Basin Hiking Area, a 15,000-acre roadless area dotted by 28 alpine lakes. Mt. Aeneas towers above these watery jewels. We only went far enough into the area to see the Picnic Lakes and two-lobed Black Lake, but if these are representative of the beauty of the area, we recommend a more thorough exploration.

If your recreation mode calls for less strenuous activity, spend your time in the Flathead around the 34-mile-long Hungry Horse Reservoir. There are many attractive campgrounds, and the fishing for trout, grayling, and whitefish is good sport. If you decide to boat, keep in mind that the maximum depth of the reservoir is 500 feet. Without moving from your campsite you can see dozens of mountain peaks.

Gallatin National Forest

Location: 1,701,000 acres in southern Montana, mostly between Bozeman and Yellowstone National Park. Interstate 90 and U.S. Highways 89, 191, 212, and 287 are the major roads. Maps and brochures available at the district ranger stations in Big Timber, Bozeman, Gardiner, West Yellowstone, and Livingston, at the Madison River Canyon Earthquake Visitor's Center, or by writing the Forest Supervisor, Federal Building, Box 130, Bozeman, MT 59715.

Facilities and Services: Campgrounds (some with camper disposal service); picnic areas with grills; boat ramps; visitor's center.

Special Attractions: Gallatin Petrified Forest; Madison River Canyon Earthquake Area; Spanish Peaks Wilderness; Absaroka-Beartooth Wilderness.

Activities: Camping, picnicking, hiking, backpacking, hunting, fishing, boating, horseback riding, winter sports, nature study.

Wildlife: Threatened bald eagle and grizzly bear; elk, Rocky Mountain bighorn sheep, black bear, mountain goat, wolverine.

ONE OF THE RAREST SITES in the United States is found in the Gallatin National Forest; it is the Gallatin Petrified Forest which covers 40 square miles. Here, standing in an upright position, are numbers of petrified or lifeless and rigid trees. Fossil experts have identified more than 100 species. Although it is permissible to collect fragments of petrified wood that are strewn over the ground, it is illegal to take specimens from the upright stumps. Even to gather fragments, you must buy a permit from a forest ranger that limits you to 25 pounds in one day or 100 pounds in one year. We began our visit from the Tom Miner Campground, a secluded spot nestled beneath pines along Tom Miner Creek west of Gardiner. To get to specimens of good upright petrified trees, you must hike nearly to Ramshorn Peak, some 4 miles distant. From the trail are rewarding views of the surrounding mountains and colorful meadow wildflowers. The sheer thrill of seeing the first upright tree is something you will not forget. These fossilized remains tell us that here is a place where time has literally stood still.

We were still talking about the excitement of seeing the petrified forest when we arrived at the Madison River Canyon Earthquake Area, another forest special attraction, outside the west entrance to Yellowstone. This area marks the spot where 28 people were buried beneath a landslide as they slept on the evening of August 17, 1959. The cause of the holocaust was a series of earthquakes with a shock estimated to be as powerful as 2,500 atomic bombs. A horrendous landslide up to 400 feet deep filled the mouth of the Madison River, and the new Earthquake Lake was formed. A visitor's center at the site explains the tragic events, and several trails and back roads into the area provide an opportunity for firsthand observation.

There are two wilderness areas in the Gallatin that are other star attractions to the forest visitor. One is the Spanish Peaks Wilderness south of Bozeman, one of the most rugged yet serene wildernesses in the United States. Snow-covered Gallatin and Wilson peaks tower like sentinels over the unspoiled land of glacial lakes, waterfalls, wildflower-filled meadows, and cool, clear trout streams. Huge Douglas firs and spruces, some reaching to 120 feet, dominate the lower slopes of the mountains and create dark, dense woods clogged with ferns and wildflowers. Around 9,000 feet, the dense forests give way to scattered, gnarly subalpine firs and whitebark pines on the windswept ridges. There are many trails to take, ranging in duration from one day to one week. Most heavily used is the trail up the South Fork of Spanish Creek to the Spanish Lakes, a distance of about eight miles with rather easy grades.

The other wilderness is the Absaroka-Beartooth Wilderness at the eastern edge of the Gallatin, shared by the Custer National Forest. Containing both the Absaroka and Beartooth mountains, this 918,762-acre region ranks among the finest in the United States for mountain beauty. The Absaroka Range, relatively young from a geological standpoint, is volcanic, while the much older Beartooth is formed of granite. Because of the longer time for erosion to take place, the Beartooths have jagged peaks, deep canyons, and sharp ridges; the Absarokas are softer and more rolling in appearance. Granite Peak, on the Gallatin-Custer boundary, is the highest peak in Montana at 12,799 feet. It is a challenge to climbers, and should be tried by only the most experienced. For those of us less experienced, we recommend trails outside the wilderness to such worthy sites as Natural Bridge Falls, Hour Glass Falls, Pine Creek Falls, or Independence, a ghost town. You will need a good forest service map to locate these choice spots.

Helena National Forest

Location: 975,000 acres in west-central Montana surrounding Helena. Interstate 15, U.S. Highway 12, and State Route 200 are the major roads to the forest. Maps and brochures available at the district ranger stations in Helena, Lincoln, and Townsend or by writing the Forest Supervisor, Federal Building, Helena, MT 59626.

Facilities and Services: Campgrounds (some with camper disposal service); picnic areas with grills; boat ramps.

Special Attractions: Gates of the Mountains Wilderness; Scapegoat Wilderness.

Activities: Camping, picnicking, hiking, backpacking, hunting, fishing, boating, horseback riding, nature study, winter sports.

Wildlife: Endangered or threatened bald eagle, American peregrine falcon, and grizzly bear; black bear, Rocky Mountain bighorn sheep, mountain goat, mountain lion, bobcat, elk, moose.

A BIG ATTRACTION in the Helena are the Gates of the Mountains and surrounding wilderness. When Meriwether Lewis entered the area he, too, was greatly impressed and recorded in his diary, "These cliffs rise from the water's edge on either side perpendicularly to the height of 1,200 feet. I called it the Gates of the Rocky Mountains." Today, unlike Lewis, you can take a commercial boat ride through the canyon on the Missouri River. At one place the angle of the limestone cliffs is such that the cliffs seem to open as the boat approaches, giving the impression of a gate. In early spring mountain goats on the cliffs are often seen watching the tourists below.

You can also drive through the surrounding and scenic Big Belt Mountains, as well. The forest service has laid out an 85-mile Figure 8 Scenic Auto Tour that skirts the edge of the Gates of the Mountains Wilderness. Heavy rains have washed out one rugged 4-mile section through Trout Creek Canyon, and the forest service has no plans at present to repair this stretch of the road. It is worth a hike, though, through Trout Creek Canyon. You can still take that part of the Figure 8 road that parallels Beaver Creek and continues to Refrigerator Canyon. A half-mile trail leads to this 15-foot-wide gorge through which cool breezes always seem to flow. You can hike beyond Refrigerator Canyon, enter the Gates of the Mountains Wilderness, and continue for 15 miles to the Meriwether Picnic Area along the banks of the Missouri River.

The Scapegoat Wilderness is the other attraction in the Helena, although only a third of it is in the forest. The remainder is in the Lolo and Lewis and Clark forests. One good hike that can easily be completed in a day is a round trip to glistening Heart Lake from the Indian Meadows Guard Station.

A few miles west of Helena, along the gravel road that parallels Sweeney Creek, is the popular Sweeney Creek Ecology Trail where you can learn some basic principles concerning the relationship of plants and animals to the environment. A guide leaflet is available that aids in identifying the dominant ponderosa and lodgepole pines and Douglas fir, the shrubby buffalo berry, birch-leaf spiraea, and snowberry, and colorful wildflowers that include the arrowleaf balsamroot. It is a good way to spend a half hour of your day.

Scattered throughout the forest are many abandoned mines, and you can visit the near–ghost town of Rimini.

Kootenai National Forest

Location: 1,778,738 acres in northwestern Montana, extending to the Canadian border. U.S. Highways 2 and 93 and State Route 37 cross the forest. Maps and brochures available at the district ranger stations in Trout Creek, Libby, Fortine, Eureka, and Troy, at the Libby Dam Visitor's Center, or by writing the Forest Supervisor, W. Highway 2, Libby, MT 59923.

Facilities and Services: Campgrounds (some with camper disposal service); picnic areas with grills; boat ramps; swimming beaches.

Special Attractions: Cabinet Mountains Wilderness; Northwest Peak Scenic Area; Ten Lakes Scenic Area.

Activities: Camping, picnicking, hiking, backpacking, hunting, fishing, boating, swimming, horseback riding, winter sports, nature study.

Wildlife: Threatened bald eagle and grizzly bear; black bear, Rocky Mountain bighorn sheep, mountain goat, elk, moose, wolverine.

THE CABINET MOUNTAINS WILDERNESS, which is shared with the Kaniksu National Forest, is a major attraction of the Kootenai, with its high, rocky peaks, densely forested ravines, deep blue lakes, and clear, cold streams, many of them forming waterfalls as they rush to their destination. Encompassing nearly 100,000 acres, this wilderness is a natural botanical garden of richly hued wildflowers and fragrant-flowering shrubs. The threatened grizzly bear roams the area, and moose and elk are often seen foraging near streams. The shaggy mountain goat is sometimes observed on the high, rocky crags.

There are several trails into the Cabinet Mountains Wilderness that can be completed in one day. We recommend the popular mile-and-a-half trail to sparkling Leigh Lake, where the fishing for trout cannot be surpassed in the forest.

Should you wish to continue beyond Leigh Lake, a more rugged trail three miles long to 9,712-foot Snowshoe Peak, the highest point in the Kootenai, on the Montana-Idaho border is worth taking. On the way, the trail swings past Bockman Peak and passes east of Blackwell Glacier. Another sterling trail into the wilderness follows Cedar Creek for five miles to two sapphire-blue lakes. En route this trail passes ancient giant western red cedars.

Another forest attraction is the Northwest Peak Scenic Area, a region of 14,000 acres of high mountain ridges in the northwest corner nearly reaching the Canadian border. This wild region, dotted by seven crystal clear lakes, is penetrated by only one trail. One treat in store for you is a splendid stand of gnarly alpine larch trees on the slopes of Northwest Peak. The trail climbs 1,000 feet in two and a half miles.

Our favorite special attraction in the Kootenai is the 18,800-acre Ten Lakes Scenic Area northeast of Eureka and pressing against the Canadian border. Glaciers have shaped the area, carving many basins that are filled with silvery lakes. The prominent ridge of the Whitefish Mountains runs the entire length of the scenic area and is a good place to look for Rocky Mountain bighorn sheep. The region is also one of the haunts for the threatened grizzly bear, so caution should be observed while hiking the area. During mid-summer, gardens of alpine flowers prevail in the meadows. As you hike in the Ten Lakes Scenic Area, you will easily see why the famous naturalist Ernest Thompson Seton was inspired to write his animal and wood lore books. Many of the lakes in the area are bordered by rugged rocky cliffs on one side and stands of whitebark pine on the other. Although the Ten Lakes Scenic Area is at high elevations that support subalpine types of vegetation, the Tobacco Valley area just six miles west stands in stark contrast. It supports cacti and other dry-inhabiting desert species.

West of the Eureka Ranger Station in the Ten Lakes Scenic Area where the Tobacco River empties into the Kootenai River is an area of fantastic rocky spires known as the Hoodoos. They have developed their weird appearance (and consequent weird name) through centuries of weathering. Other entertaining rock spires in even more rugged terrain are found near the Bad Medicine Campground at the southern end of Bull Lake. We found this a great place to spend a few nights because of the engaging scenery that surrounds the campsites. For more secluded camping, we recommend the Yaak Falls Campground, a short walk from the thundering falls. More accessible are the Kootenai Falls along U.S. Highway 2 west of Libby. There the Kootenai River spills over a series of low ledges, but the width of the river falls makes up for the short cascades.

Although many giant red cedars grow in the Kootenai, the most impressive grove is found along Ross Creek. Take the mile-long nature trail that winds through the 100-acre grove. You will pass trees more than 500 years old, up to 175 feet tall, and nearly 8 feet in diameter. After enjoying that outing, if you are tempted to fish for cutthroat trout in cool, clear Ross Creek, you can fry and savor your catch in the nearby picnic area.

Lewis and Clark National Forest

Location: Almost two million acres in west-central Montana, west and southeast of Great Falls. U.S. Highways 2 and 89 are the primary roads. Maps and brochures available at the district ranger stations in Choteau, Stanford, Harlowton, and White Sulphur Springs or by writing the Forest Supervisor, 1601 2nd Ave. N, Great Falls, MT 59403.

Facilities and Services: Campgrounds (some with camper disposal service); picnic areas with grills; boat ramps; swimming beaches.

Special Attractions: Chinese Wall in the Bob Marshall Wilderness; Scapegoat Wilderness.

Activities: Camping, picnicking, hiking, backpacking, hunting, fishing, boating, swimming, floating, horseback riding, nature study, winter sports.

Wildlife: Endangered or threatened grizzly bear, bald eagle, and American peregrine falcon; black bear, Rocky Mountain bighorn sheep, mountain goat, mountain lion, bobcat, elk, moose.

THERE ARE TWO DISTINCT SECTIONS of the Lewis and Clark National Forest. One section is west of Great Falls, Montana, extending south from Glacier National Park. This is the Rocky Mountain section, where mountain peaks, spectacular steep ridges, narrow valleys, and glacial lakes abound. The other section, east and southeast of Great Falls, boasts of six distinct mountain ranges but most of them are relatively gentle, with rounded summits and deep valleys.

As you might suspect, the Rocky Mountain section is the wildest; in it are two designated wilderness areas, the Bob Marshall and the Scapegoat, both forest attractions. Half of the Bob Marshall is in the Flathead National Forest and discussed under that entry. The Lewis and Clark share of this wilderness is a formidable area, highlighted by the incomparable Chinese Wall, a 15-mile-long, straight-sided escarpment of limestone that rises 1,000 feet above the valley floor. We had seen several pictures of the Chinese Wall, and looked forward to it with eager anticipation. In spite of being primed for the event, our first view was more glorious than ever anticipated. The vertical cliffs do rise to a mighty 1,000 feet, and although we have yet to see the Great Wall of China, we knew we were facing a great wall—but one made by nature. It is not an easy journey to get to it. The best route, which is close to 30 miles one way by trail, is via Headquarters Creek Pass. As one drives the narrow road to the Headquarters trail head, powerful Mill Falls provides a splashing welcome near the gravel lane. Once over the pass, it is a matter of following the trail to the Gates Park Guard Station and Rock Creek until you reach the wall.

We next visited the Scapegoat Wilderness and were amazed to see a similar but smaller vertical limestone wall on the east face of Scapegoat Mountain. This wall is of easier access, being only a few miles from the Benchmark Road that snakes its way up Wood Canyon. Before leaving the road for the wall, stop and view the unique Double Falls of Ford Creek, a short way off the road.

The less rugged section of the forest east and southeast of Great Falls is called the Jefferson division. Of its six ranges, Big Snowy, Little Snowy, and Castle Mountains are composed of limestone, the Crazy and Highwood mountains are volcanic, and the Little Belt Mountains are mostly sandstone and shale. Big Snowy has a fascinating ice cave which is so cold that it maintains perpetual ice within. The cave is two miles south of Crystal Lake with the bubbling white-water Crystal Cascades between. Little Belt Mountains contains Judith Basin, which was often depicted in the work of noted Western artist Charles M. Russell. Many abandoned mines in the Little Belt, particularly near Neihart, and several abandoned towns, such as Old Yogo, are interesting to explore.

Lolo National Forest

Location: 2,062,545 acres in western Montana, surrounding Missoula. Interstate 90 and State Routes 135 and 200 are the major routes. Maps and brochures available at the district ranger stations in Missoula, Huson, Plains, Seeley Lake, Superior, and Thompson Falls, at the Lolo Pass Visitor's Center, or by writing the Forest Supervisor, Building 24, Ft. Missoula, MT 59801.

Facilities and Services: Campgrounds (some with camper disposal service); picnic areas with grills; boat ramps; swimming beaches.

Special Attractions: Welcome Creek Wilderness; Rattlesnake Wilderness; Scapegoat Wilderness; Selway-Bitterroot Wilderness.

Activities: Camping, picnicking, hiking, backpacking, hunting, fishing, boating, swimming, horseback riding, nature study, winter sports.

Wildlife: Endangered or threatened grizzly bear, Rocky Mountain wolf, bald eagle, and American peregrine falcon; elk, moose, Rocky Mountain bighorn sheep, mountain goat, black bear, wolverine, Canada lynx, mountain lion, bobcat.

THE LOLO has it all—snow-bonneted mountain peaks, sweeping streams, splashing waterfalls, meadows of brightly colored wildflowers, wild backcountry, historic buildings, and wildlife. The wildlife is particularly outstanding, with federally listed endangered or threatened animals inhabiting the forest. It is not uncommon to see the stalwart bald eagle and American peregrine falcon flying overhead, but the grizzly bears and the Rocky Mountain wolves are rarely seen. If you do come upon them, make every effort not to provoke an encounter.

The forest's four wilderness areas get top billing on a list of things to do and see here. One of them is a 28,000-acre wilderness in the Sapphire Mountains known as the Welcome Creek Wilderness. There is easy access to it from several campgrounds along the way via Forest Road 102. Although there are scenic mountain vistas such as 7,723-foot Welcome Peak and 7,046-foot Cinnamon Bear Point, we actually found the remains of several old mines more fun and poked around them.

Another small wilderness is the 33,000-acre Rattlesnake Wilderness. Because the area is only about seven miles from the city of Missoula, it is not a wilderness area where solitude prevails. Nonetheless, the steep-sided narrow valleys and glacier-formed lakes provide adequate outdoor recreation for everyone. Forest Trail 517 to Stuart Peak is popular, but if you have time, continue on to Mosquito Peak with trout-filled Glacier Lake to the north.

A portion of the ruggedly scenic Scapegoat Wilderness is also in the Lolo. There are trails that follow picturesque creeks such as Canyon, Dobrota, Dwight, and Cooney, as well as the Blackfoot River. Only a small part of the Selway-Bitterroot Wilderness is in the Lolo. See entry for the Bitterroot National Forest.

We also explored the Bitterroot Mountains on the western edge of the Lolo. There are several back roads that lead southwest from Interstate 10 between Lookout Pass and Superior toward such landmarks as Dominican Peak, Ward Peak, Flattop Mountain, Black Peak, and Little Joe Mountain. We decided to head for Diamond Lake Campground and took Forest Road 342 west of Superior. There are many lovely and secluded sites on which to camp along sparkling Diamond Lake. From here trails of less than five miles wind and climb to rocky Torino Peak and rugged Eagle Cliff.

Another scenic trip into the Bitterroots follows Forest Highway 7 from Thompson Falls up and over Thompson Pass and finally into the Coeur d'Alene Forest. As the road approaches the Bitterroots, plan to stop for a while at the Mountain House and Halfway House. They are historic sites and are National Historic Places.

There are also two short hikes that we suggest in the

Lolo. One is the one-and-a-half-mile round trip to Cascade Falls, a gushing waterfall on Cascade Creek. Along the way are a couple of sweeping views of the Clark Fork River Valley. The other trail is the three-and-a-half-mile loop to Skookum Butte. At its lookout spot is a full circle overview that includes the Selway-Bitterroot and Rattlesnake wildernesses.

Black Hills National Forest

Location: 1,233,000 acres in southwestern South Dakota and adjacent Wyoming, west of Rapid City and surrounding Mt. Rushmore National Monument. Interstate 90, U.S. Highways 16, 16A, 85, and 385, and State Routes 44, 71, 89, and 24 cross the forest. Maps and brochures available at the district ranger stations in Custer, Hill City, Deadwood, Rapid City, and Spearfish, South Dakota, and Sundance and Newcastle, Wyoming, at the Pactola Reservoir Visitor's Center, or by writing the Forest Supervisor, Box 792, Custer, SD 57730.

Facilities and Services: Campgrounds (some with camper disposal service); picnic areas with grills; boat ramps; swimming beaches; visitor's center.

Special Attraction: Black Elk Wilderness.

Activities: Camping, picnicking, hiking, backpacking, hunting, fishing, swimming, boating, waterskiing, horseback riding, nature study, winter sports.

Wildlife: Threatened bald eagle; mountain goat, mountain lion, Rocky Mountain bighorn sheep.

THE BEAUTIFUL BLACK HILLS rise abruptly at the northern end of the Great Plains. They cover an area roughly 110 miles long and 70 miles wide in South Dakota and adjacent Wyoming. Most of the area is incorporated in the Black Hills National Forest.

The most rugged features are in the center of the forest surrounding Mt. Rushmore National Monument in the Black Elk Wilderness. The highest and most impressive point is Harney Peak, whose broad and massive granite summit at 7,242 feet is the highest elevation east of the Rocky Mountains and a focal point of the wilderness. This is a land of heavily forested ridges, rugged peaks, and shaded valleys that harbor a wide diversity of plant and animal life. Most of the forests are dominated by ponderosa pine, but there are several isolated occurrences of plants that are unexpected in South Dakota and always fascinate naturalists. For example, there is an isolated 90-acre stand of lodgepole pine that is more than 200 miles from the nearest stand in the Big Horn Mountains of Wyoming. In the vicinity of some of the granite spires is a small stand of limber pine, also displaced from the Big Horns. Botanists have recorded nearly two dozen "unexpected" plants, including maidenhair fern, twinleaf, wild wintergreen, maidenhair spleenwort, and the tiny moschatel. Some of these plants are more typical to the East, some to the North, and some to the West.

Harney Peak is also home for the magnificent mountain goat, as well as mountain sheep and mountain lion. Songbirds, too, are plentiful. No matter where you decide to hike in the vicinity of Harney Peak and the Black Elk Wilderness, you will be rewarded by beautiful forests, interesting rock formations, and a variety of living things.

Northwest of Harney Peak, the Black Hills take on a different aspect with limestone canyons, waterfalls, and even a high peak or two such as Terry Peak, which reaches 7,071 feet. Spearfish Canyon is one of the loveliest in the forest and is traversed by U.S. Highway 85. North of here is the ruggedly scenic Roughlock Falls whose several cascades cause a swirling turbulence as they drop into a waiting stream.

There are major recreation areas developed around Pactola Reservoir and Sheridan and Deerfield lakes. All are nestled in a superb mountain setting and have several campgrounds to choose from. There is a visitor's center at the eastern end of Pactola.

Any visitor to Mt. Rushmore National Monument must pass through the Black Hills National Forest. The Iron Mountain Road (U.S. Highway 16A) from the south is the most spectacular with its "pig-tail switchbacks," tunnel, and the Norbeck Memorial Overlook. Stop at the overlook for a breathtaking view of the Mt. Rushmore stone carvings in the distance.

Most of the Wyoming portion of the Black Hills lies north

of Sundance in the Bear Lodge Mountains. The scenery is particularly dazzling in the vicinity of the Warren Peaks.

Big Horn National Forest

Location: 1,113,700 acres in northern Wyoming, west of Sheridan and Buffalo. U.S. Routes 14, 14A, and 16 are the major roads. Maps and brochures available at the district ranger stations in Buffalo, Lovell, Greybull, Worland, and Sheridan or by writing the Forest Supervisor, 1969 S. Sheridan Ave., Sheridan, WY 82801.

Facilities and Services: Campgrounds (some with camper disposal service); picnic areas with grills.

Special Attractions: Medicine Wheel Archeological Site; Cloud Peak Primitive Area.

Activities: Camping, picnicking, hiking, backpacking, hunting, fishing, horseback riding, winter sports, nature study.

Wildlife: Rocky Mountain bighorn sheep, black bear, elk, moose, golden eagle.

HIGH ON A SECLUDED PEAK of the Big Horn Mountains is the mysterious Medicine Wheel, the star attraction of the Big Horn National Forest. The wheel is a series of side-by-side stones forming a perfect circle with a diameter of 80 feet. Situated around the perimeter are six piles of stones two and a half feet tall. The center of the wheel contains a hub with 28 radiating spokes. It is speculated that early Indians used the wheel as a primitive instrument to set time for ritual ceremonies.

The road to the Medicine Wheel is treacherously narrow. We stopped a couple of times to photograph the lovely dwarf alpine flowers, including a lavender shooting star. A golden eagle soaring overhead kept an eye on us.

The wildest part of the Big Horn National Forest has been set aside as the Cloud Peak Primitive Area. The 137,000 acres are at elevations of between 8,500 and 13,165 feet. Sky blue Lake Solitude, surrounded by rough, craggy cliffs, is the most popular feature and is reached by a 10-mile trail from the Paintrock Guard Station. We recommend the more accessible string of lakes that comprise the Seven Brothers Lakes.

We ran across a peaceful area several miles northeast of the Medicine Wheel called Bull Elk Park. The park is a treeless, grassy bald rimmed by a stand of lodgepole pine. The bald is a mixture of western grasses and colorful wildflowers. During the summer, blue lupines and yellow cinquefoils offer bright contrasts to the Idaho fescue grass.

There are three major highways that have dramatic entrances into the Big Horn. Route 16 from Worland enters the forest via Tensleep Canyon, a rocky gorge carved by Tensleep Creek. East from the village of Shell on Route 14 is Shell Canyon. After a few switchbacks, the highway comes to the Post Creek Picnic Area, a secluded spot along a tributary that feeds into the canyon. The picnic area is at the western edge of a superior example of an open woods, dominated by the Rocky Mountain juniper.

The old section of Route 14A from Lovell climbs to the top of the Big Horns on the longest, curviest mountain road we have ever driven. If you like the thrill of a hairpin-curving mountain road, you will not want to miss this one. If you are squeamish, you can bypass the climb by staying on the newly opened section of the highway.

Bridger-Teton National Forest

Location: 3,440,000 acres in western Wyoming, east and south of Grand Teton National Park. U.S. Highways 26, 187, 189, and 287 and State Route 22 are the major highways. Maps and brochures available at the district ranger stations in Big Piney, Moran, Afton, Jackson, Kemmerer, and Pinedale or by writing the Forest Supervisor, 340 N. Cache, Jackson, WY 83001.

Facilities and Services: Campgrounds (some with camper disposal service); picnic areas with grills; boat ramps; swimming beaches; horse corral and loading ramp.

Special Attractions: Bridger Wilderness; Kendall Warm Springs; Periodic Springs; Gros Ventre Slide Area; Teton Wilderness.

Activities: Camping, picnicking, hiking, backpacking, hunting, fishing, boating, swimming, horseback riding, nature study, winter sports.

Wildlife: Endangered or threatened bald eagle, rare trumpeter swan, grizzly bear, and Kendall Warm Springs dace; moose, elk, Rocky Mountain bighorn sheep, mountain lion, bobcat.

THE BRIDGER WILDERNESS, a prime forest attraction, makes up most of the eastern part of the forest. It is a land of exceptionally rugged terrain punctuated by more than 1,000 beautiful lakes. Several glaciers hang near the top of the higher peaks. The 13,785-foot Gannett Peak, on the Continental Divide which separates the Bridger-Teton from the Shoshone National Forest, is one of 20 peaks in the wilderness above 12,000 feet. In all, there are more than 500 miles of trails. One great hike is from the trail head at Green River Lakes past Square Top Mountain and Granite Peak to the Green River Pass and Summit Lake. When we took the trail, we were rewarded with a glimpse of a moose with her yearling offspring and another glimpse of a large elk. Excellent campsites are at New Fork Lakes, Half Moon Lake, and the large Fremont Lake.

On the way to Green River Lakes is Kendall Warm Springs, another forest attraction. The springs have a constant 84-degree water temperature, and they are the only place in the world where the Kendall Warm Springs dace, a two-inch-long fish, breeds. The males are purple and the females green at breeding time. It is an endangered species. The stream formed by the springs rushes over a small cliff and into the Green River. A short distance north on the side of a hill are the dilapidated cabins and ruins of the Billy Wells Dude Ranch, reputedly the first dude ranch in the country.

South of the Snake River Canyon in the Middle Range is the southern part of the Bridger division. Five miles east of Afton is another special forest sight, Periodic Springs, where a trickle of cold water gradually builds up to a turbulent flow before completely shutting down after 18 minutes, only to begin the process over again. This regular rhythmic type of spring is said to be found elsewhere only in New Zealand, Yugoslavia, and Jerusalem. The road and trail to the springs are in a scenic canyon through which the Swift River rushes. Interesting rock formations including a towering balanced rock are along the way. At one point water pours from an opening in a cliff and cascades 50 feet below to the Swift River.

The Teton division of the national forest lies north, east, and south of Jackson. There are several geological highlights in this section. Most startling and famous is the Gros Ventre Slide Area where, on June 23, 1925, 50 million cubic yards of sandstone slid off the mountain and across the Gros Ventre River, damming it up and forming Slide Lake and leaving a scar one mile long, 2,000 feet wide, and a few hundred feet deep in the side of the mountain. A half-mile trail leads to several vantage points in the area.

Along picturesque Granite Creek north of U.S. Highway 189 is Granite Hot Springs, a thermal pool suitable for swimming. The nearby Granite Creek Campground is set among spectacular scenery and is only a short distance away from Granite Falls.

Two other waterfalls are in the Teton Wilderness. The falls on the South Fork of the Buffalo River drop nearly 100 feet into a narrow canyon, while the North Fork Falls drop several times in a rough, rocky gorge.

The Teton Wilderness itself, another drawing card of the forest, is 563,500 acres of mountain peaks, rocky canyons, and high plateaus along the Continental Divide and adjacent to the southern boundary of Yellowstone National Park. Unusual rough and broken rock formations composed of breccia, a type of cemented rock, are found in the northeast section. A dramatic example of this are the Breccia Cliffs, a short distance north of Togwotee Pass. At Two Ocean Pass on the Continental Divide, Two Ocean Creek abruptly splits into Atlantic and Pacific creeks, one that heads toward the Atlantic Ocean, the other toward the Pacific. An excellent long loop trail leaves from the Turpin River Campground, continues along the South Fork of the Buffalo River, and leads to the South Fork Falls, through lovely wildflower-laden Pendergraft Meadows, and along the Yellowstone River through seven-mile-long Yellowstone Meadows to Bridger Lake, a good place to see the rare and graceful trumpeter swan.

Medicine Bow National Forest

Location: 1,093,000 acres in southeastern Wyoming, north and west of Cheyenne. Interstates 25 and 80 and State Routes 130 and 230 are the access routes. Maps and brochures available at the district ranger stations in Saratoga, Encampment, Laramie, and Douglas or by writing the Forest Supervisor, 605 Skyline Drive, Laramie, WY 82070.

Facilities and Services: Campgrounds (some with camper disposal service); picnic areas with grills; boat launching areas; swimming beaches.

Special Attractions: Snowy Mountain Range; Vedauwoo Rocks; Savage Run Wilderness.

Activities: Camping, picnicking, hiking, backpacking, hunting, fishing, boating, swimming, horseback riding, winter sports.

Wildlife: Rocky Mountain bighorn sheep, elk, moose, mountain lion, bobcat.

ON MEMORIAL DAY snowplows cleared State Highway 130, which lies across the forest's Snowy Mountain Range west of Laramie. The range is one of forest's major attractions. The next day we left Centennial and climbed rapidly to the 10,847-foot Snowy Range Pass, then descended just as quickly down the western slope before finally reaching Saratoga. It was reminiscent of a toboggan ride, with the snowbanks piled as high as 15 feet on both sides of the road. The snow piles curtailed any activities off the roadway, but the scenery was breathtakingly spectacular. When we returned to the area in August, the only snow left was on the high peaks; the many sapphire-blue alpine lakes that earlier were obscured by snow were now sparkling like so many precious blue jewels. Our memories will never fade of Lake Marie, Mirror Lake, and Lookout Lake in the shadows of mighty Medicine Bow Peak. Near the pass is Libby Flats Observation Site where there are unexcelled views of the Snowy Mountain Range. There are hiking trails from here that pass many of the hundred alpine lakes.

Southwest of the Snowy Mountain Range and west of the community of Encampment is the Sierra Madre Range which straddles the Continental Divide. The road across the divide from Encampment to Boggs, Colorado, exceptionally beautiful in the autumn when the aspens turn color, passes the site of the old mining town of Battle. A good trail from Battle leads to the 11,000-foot Bridger Peak some four miles away. Another scenic trail in this area follows the Encampment River through the rugged Encampment River Canyon to the Hog Park Reservoir. The Lakeview Campground at the west end of the reservoir is a good place to watch the setting sun.

Rising from the Great Plains at the eastern edge of the forest are the rugged Laramie Mountains, a part of the Front Range of the Rockies. There are three choice trails in these mountains. One is five and a half miles to the top of Laramie Peak from the Friend Park Campground. From the summit of Laramie Peak is a view that stretches all the way to Scottsbluff, Nebraska, to the east, and Laramie, Wyoming, to the southwest. Another scenic trail is through the granitic La Bonte Canyon, a good area to look for bighorn sheep. Finally, the Twin Peaks Trail near La Prele leads through an area of interesting rock formations, including a double-pointed mountain.

No visit to the Medicine Bow is complete without a detour to the Vedauwoo Rocks. These remarkable rock formations are found between Laramie and Cheyenne in the Sherman Mountains. The granite boulders take many strange shapes. One particularly fine concentration is known as the Devil's Playground; another well-named rock is Turtle Rock. At Vedauwoo Glen, you will find a two-stage theater carved into a solid wall of granite. For a further bonus, there is a lovely picnic area in the midst of the Vedauwoo; you will have an entertaining lunch surrounded by all these curious formations.

A final "must" stop for visitors is the Savage Run Wilderness, a 14,000-acre area of virgin timber. The region is interrupted occasionally by grassy slopes that teem with brightly colored summer wildflowers. Sagebrush communities are scattered here and there. The principle trees are Engelmann spruce, Douglas fir, subalpine fir, and lodgepole pine. To add to your enjoyment there are a few hiking trails, although horseback riding is popular from the nearby A Bar A Ranch, which rents horses.

Shoshone National Forest

Location: 2,433,125 acres in northern Wyoming, east of Yellowstone National Park. U.S. Highways 14, 16, 20, 26, 212, and 287 are the major roads. Maps and brochures available at the district ranger stations in Powell, Meeteetse, Lander, Cody, and Dubois or by writing the Forest Supervisor, 225 West Yellowstone Highway, Cody, WY 82414.

Facilities and Services: Campgrounds (some with camper disposal service); picnic areas with grills; boat ramps.

Special Attractions: North Absaroka Wilderness; Washakie Wilderness; Fitzpatrick Wilderness; Popo Agie Primitive Area; Whiskey Mountain Primitive Area.

Activities: Camping, picnicking, hiking, backpacking, hunting, fishing, boating, horseback riding.

Wildlife: Threatened bald eagle and grizzly bear; black bear, Rocky Mountain bighorn sheep, mountain goat, moose, elk, mountain lion, bison, wolverine, trumpeter swan.

ONE OF THE BEST WAYS to get an overview of the wild and rugged Shoshone National Forest is to drive the spectacular Beartooth Highway from Red Lodge, Wyoming, to Cooke City, Montana. The highway climbs dramatically out of Red Lodge to the 10,947-foot summit at Beartooth Pass. A stop at the pass itself, well above timberline, is an alpine experience. The ground is covered with pebbles, among which peak a myriad of dwarf, colorful, tundralike wildflowers. Pink primroses, lavender shooting stars, and blue gentians are just a few of the delicate beauties that bloom from mid-June to late August. Other short side trips from the Beartooth Highway include Island, Beartooth, and Lily lakes, three shimmering pools where fishing for trout reportedly is excellent. Just before the Beartooth Highway leaves the Shoshone and enters Montana, it comes nearly face-to-face with spire-pointed Pilot Peak, a well-known landmark that reaches 11,708 feet. The prominent mountain north of Pilot Peak is the less spectacular Index Peak.

Backcountry enthusiasts may choose from three wilderness areas and two primitive areas—all of which are star attractions of the forest. The 350,000-acre North Absaroka Wilderness, which extends to Yellowstone's eastern border, is a wildlife haven. Elk, moose, black bear, and bison are regularly seen here, while less often observed but still present are the grizzly bear, lynx, bobcat, and Rocky Mountain bighorn sheep. The Papoose Trail is a long and strenuous route through the wilderness from the Crandall Forest Service Station in the Shoshone to the Upper Miller Creek Patrol Station in Yellowstone. If you plan to hike it, allow at least three days one way.

Twice as large and equally as wild is the Washakie Wilderness which lies south of the Cody-Yellowstone Highway. Among the interesting attractions in the Washakie is Blackwater Natural Bridge, an impressive stone arch nearly 10 miles by trail from the Rex Hale Picnic Grounds.

The Fitzpatrick Wilderness south of the Wind River contains the roughest terrain in the Shoshone. The Continental Divide, that imaginary line that separates the water flow to the Atlantic and the Pacific oceans, forms the western boundary of the wilderness and is dominated by several peaks in excess of 13,500 feet. Tallest is 13,804-foot Gannett Peak, the highest mountain in Wyoming. Several large active glaciers cover some of the upper slopes; most prominant are Downs, Gannett, Dinwoody, Fremont, and Bull Lakes glaciers. Hundreds of high mountain lakes add variety to the wilderness.

Another high mountain, roadless area with jagged peaks, deep canyons, and alpine meadows is the 71,000-acre Popo Agie (pronounced po-po-shee-ah) Primitive Area west of Lander. Also on the Continental Divide, this rugged area is not open until mid-July because of the heavy snow.

The last forest star attraction is five miles south of Dubois and is the Whiskey Mountain Primitive Area. This land has been set aside to preserve the largest herd of Rocky Mountain bighorn sheep in the United States. Forest visitors may enter and enjoy the area.

The Cody-Yellowstone Highway follows the North Fork of the Shoshone River through picturesque Wapiti Valley. You can choose from among a dozen campsites situated along the river. Several unusual rock formations loom over the valley, including the well-named Elephant Rock and the Holy City, where the rocks give the appearance of an abandoned medieval city.

An exciting drive over a good road through the backcountry is Wyoming Route 296. It leaves the Beartooth Highway near the Lake Creek Campground and follows the rushing Clarks Fork of the Yellowstone River as it winds through the Absaroka Mountains. The excitement comes when you negotiate the series of sharp switchbacks leading to Dead Indian Pass. There are several suitable campgrounds along the way.

Boise National Forest

Location: 2,600,000 acres in south-central Idaho, north and northeast of Boise. Interstate 84, U.S. Highway 20, and State Routes 21 and 55 are the major routes. Maps and brochures available at the district ranger stations in Boise, Cascade, Emmett, Idaho City, and Mountain Home or by writing the Forest Supervisor, 1750 Front Street, Boise, ID 83702.

Facilities and Services: Campgrounds (some with camper disposal service); picnic areas with grills; boat ramps; swimming beaches.

Special Attractions: Sawtooth Wilderness; River of No Return Wilderness.

Activities: Camping, picnicking, hiking, backpacking, hunting, fishing, boating, swimming, horseback riding, nature study, winter sports.

Wildlife: Endangered and threatened bald eagle, American peregrine falcon, and northern Rocky Mountain wolf; rare wolverine, Canada lynx.

THE BOISE NATIONAL FOREST, about 180 miles long and 70 miles wide, has a variety of features to sample. It contains rugged mountains, high mountain meadows, deep canyons, rippling mountain streams, postcard-pretty lakes, and desertlike conditions in some of the valleys.

There are several highlights that the Boise shares with other national forests. One is the Sawtooth Wilderness, much of which is in the Boise. This is a wild region of spike-peaked mountains and soothing lakes. Ardeth and Spangle are two of the loveliest lakes in this section.

South of the Sawtooth Wilderness is the abandoned mining town of Atlanta. From the Power Plant Campground, explore around the old town and then venture north into the wilderness area.

Another highlight and shared wilderness is the River of No Return Wilderness. The wild Middle Fork of the Salmon River flows through this section separating the Boise from the Challis National Forest. Big Baldy, a granite-peaked monolith, is also here. From its summit you can enjoy spectacular vistas in all directions.

There are several man-made reservoirs in or bordering the Boise. The forest service has camping and boating facilities at Cascade, Sagehen, and Deadwood reservoirs. We like the Sagehen Reservoir in the autumn because it is then that the aspens are transformed into golden-leafed trees.

To get a complete cross section of the forest, drive from Boise diagonally across the forest to Dagger Falls at the lower end of the Middle Fork of the Salmon River. The road follows Moores Creek through Boise Basin and then between Pilot Peak and Sunset Peak to Lowman, where the route crosses the scenic South Fork of the Payette River. Above the river the road follows the course of Clear Creek before it wriggles along Bear Valley Creek to Bruce Meadows. It is then only a few miles along Dagger Creek to Dagger Falls, an entrancing 15-foot cascade. There are several pleasant campgrounds all along this route.

You might like to visit several of the hot springs in the Boise; the most popular is Kirkham Hot Springs along the South Fork of the Payette River east of Lowman. A most enjoyable trail follows the scenic South Fork of the Payette from Lowman to Banks. At times the trail veers along the river; at other times it climbs upward above the tormented water.

Caribou National Forest

Location: 1,082,368 acres in southeastern Idaho and a small part of northern Utah. Interstate 15, U.S. Highways

30, 89, and 91, and State Route 36 are the major accesses. Maps and brochures available at the district ranger stations in Malad, Montpelier, Pocatello, and Soda Springs or by writing the Forest Supervisor, 250 S. 4th Avenue, Pocatello, ID 83201.

Facilities and Services: Campgrounds (some with camper disposal service); picnic areas with grills; boat ramps; swimming beaches.

Special Attraction: Minnetonka Cave.

Activities: Camping, picnicking, hiking, hunting, fishing, swimming, nature study, horseback riding, winter sports.

Wildlife: Bear, moose, elk, mountain lion, bobcat.

CARIBOU GIVES its Minnetonka Cave top billing as a must for visitors. And well it might. The cave extends for a half mile and has many beautiful stalactite and stalagmite formations. The forest service has installed a lighting system so that some of the more outstanding formations, such as the Three Sisters and Slab of Bacon (both of which they resemble), can be enjoyed. Rangers lead guided tours every hour from 10 to 5 during the summer. The cave is located at the end of the road leading to St. Charles Canyon. For the more hardy, a steep two-mile trail leads from near the Porcupine Campground to the cavern entrance at 7,700 feet. The climb is strenuous.

In adjacent Bloomington Canyon is 10-acre, glacier-fed Bloomington Lake, which is 263 feet deep. One of the largest Engelmann spruces in the world is nearby as is an ice cave that is so cold it has ice inside year-round. It is near the end of the Paris Canyon Road.

South of Pocatello are prominent Scout and Tom mountains. A most scenic campground beneath huge Douglas firs is on the lower slope of Scout Mountain. The campground contains a natural, bowl-shaped amphitheater with terraced stone benches. The route from Pocatello to the Scout Mountain Campground passes several interesting points. One is the Cherry Springs Nature Area, a wet ravine, where vegetation becomes junglelike during the summer. Along the stream are narrow-leaf cottonwood, bigtooth maple, box elder, willow, water birch, and chokecherry. During June the chokecherries are in full white bloom and beautify and perfume the entire ravine. As the road begins its climb to Scout Mountain, a sign points down a narrow dirt trail to a Cattleman's Association Cabin. At the end of this quarter-mile lane is an old one-room log cabin still used in summer by a cattle rancher through a lease arrangement with the forest service. A little farther along is the trail head for the Crestline Cycle Trail, an eight-mile loop around Scout Mountain that is also a good hiking trail. Large Douglas firs tower over the trail head, and several stone foundations, remnants of Civilian Conservation Corps (CCC) days in the 1930s, add a historical touch to the area.

At the far eastern edge of the Caribou is the huge Palisades Reservoir. The Caribou maintains several recreation areas on the western shore where all types of water activities are possible.

We also liked the string of three Swan Lakes in the Aspen Range southeast of Soda Springs. A half-mile trail from the end of a gravel road east of U.S. Highway 30 leads through aspens and firs to the lakes.

Challis National Forest

Location: 2,463,764 acres in central Idaho, mostly between Salmon and Stanley. U.S. Highway 93 and State Route 25 are the major access routes. Maps and brochures available at the district ranger station in Challis, Mackay, and Clayton or by writing the Forest Supervisor, Forest Service Building, Challis, ID 83226.

Facilities and Services: Campgrounds (some with camper disposal service); picnic areas with grills.

Special Attractions: Sawtooth National Recreation Area; Sawtooth Wilderness; River of No Return Wilderness.

Activities: Camping, picnicking, hiking, backpacking, hunting, fishing, horseback riding.

Wildlife: Mountain goat, Rocky Mountain bighorn sheep, bobcat, mountain lion.

THE CHALLIS is a land of tall mountains and deep gorges, of alpine lakes and glacier-carved basins, of abandoned mines and ghost towns. Borah Peak, at 12,655 feet, is the tallest in the forest and in the state of Idaho. Good hikers without mountain climbing experience can reach the summit and back in one day. There is a mighty view from the top.

Our favorite mountains in the Challis—and a must special feature to see—are the beautiful White Cloud Peaks and spectacular Castle Peak, in the Sawtooth National Recreation Area. Good hiking trails lead into the White Clouds from the end of Boulder Creek Road and the Fourth of July Creek Road in the Sawtooth.

Only a very small part of the Sawtooth Wilderness is in the Challis, but enough to contain several mountain lakes, including Sawtooth Lake. With Mt. Regan as a backdrop, this lake is especially pretty. A trail from Iron Creek Campground leads to the lake and its small companion, Alpine Lake.

A portion of the River of No Return Wilderness is a further special forest attraction. The Middle Fork of the Salmon River runs through this area and separates the northern and northwestern parts of the Challis from the Boise and Salmon forests. Floating is the only way to follow this untamed river, and that should only be done in the company of commercial outfitters in the area.

For a fascinating hike or drive, leave the community of Challis by following Garden Creek on Forest Road 070. After the creek swings south, stay on the road to the site of an old toll gate through which early travelers had to pass. Beyond the toll gate site are what is left of Custer and Bonanza, two once-booming gold-mining towns. The old schoolhouse still stands at Custer and houses a museum of local memorabilia which is open during the summer months. A saloon and a few old cabins, all in good repair, are also at Custer. The buildings at Bonanza are badly crumpled, but the Bonanza Cemetery, west of the ghost town, still has tombstones, many of which are marked "Unknown." The Yankee Fork River that follows the road from Bonanza to Sunbeam Hot Springs is one of the wildest white-water streams in central Idaho. South of Bonanza, near the junction of the Yankee Fork with the Salmon River, is Sunbeam Hot Springs, one of several thermal springs in the Challis.

Hunting for agates and other rocks is good in the Challis, particularly along Morgan Creek in the northern tip of the forest. Geodes are present, also, and opals are said to be here.

One of the most interesting and least visited parts of the Challis is Pass Creek Canyon north of Leslie, where turbulent Pass Creek has gouged a gorge through a rough granite mountain. There are several caves in the cliffs, including Hidden Mouth, reached by a trail from the gravel road that twists through the canyon.

Clearwater National Forest

Location: 1,700,000 acres in northeastern Idaho. U.S. Highway 12 is the major route in the forest. Maps and brochures available at the district stations in Orofino, Kooskia, Potlach, and Kamiah, Idaho, and Lolo, Montana, at the Lolo Pass Visitor's Center, or by writing the Forest Supervisor, Rt. 1, Orofino, ID 83544.

Facilities and Services: Campgrounds (some with camper disposal service); picnic areas with grills.

Special Attractions: Historic Lolo Trail; Selway-Bitterroot Wilderness; Mallard-Larkins Pioneer Area.

Activities: Camping, picnicking, hiking, backpacking, hunting, fishing, boating, horseback riding, nature study.

Wildlife: Black bear, elk, moose, bobcat, mountain lion, Rocky Mountain bighorn sheep, mountain goat.

WHEN YOU FOLLOW the one-lane Lolo Trail through the Clearwater National Forest, you can almost sense the presence of the Nezperce Indians who used this same route many years ago to cross the Bitterroot Mountains on their way to the buffalo hunting grounds. Lewis and Clark also used this trail in their 1805 expedition seeking a route to the Pacific Ocean. Later, in 1877, Chief Joseph led about 700 of his Nezperce tribe over approximately the same route with the United States Cavalry in hot pursuit. This historic trail is approximately 150 miles long and crosses both the Clearwater and Lolo national forests. No matter where you enter the trail, you are sure to find something that attracts the eye, such as a rock formation known as the Devil's Chair, or two piles of stone called the Indian Post Office, reputed to be trail markers used by the Indians, or an Indian grave near Indian Grave Peak where a mass slaughter occurred in 1893.

If you do not wish to follow the primitive Lolo Trail, you can drive modern U.S. Highway 12, the Lewis and Clark Trail a few miles south, which more or less parallels the Lolo. Many attractions in the Clearwater are readily accessible from this highway. The first time we drove this route several years ago, we saw two black bears browsing in the vegetation a few yards from the road's edge. As you proceed a few miles from Lowell toward Lolo Pass, you will come to the Major Fenn Picnic Area where there is a National Recreation Trail that winds for a half mile through a plant community more typical of the western Cascades than this eastern part of the system. The western sword fern, red osier dogwood, and giant horsetail are just a few of the plants one can spot here. Nearby, the radiant, white-flowered Pacific dogwood, rare east of the Cascades, can be found. Four miles beyond the Major Fenn area, the highway bends north into Black Canyon. A trail (#257) leads west for about four miles to misty Smoky Peak and wildflower-laden Bimerick Meadow. A short distance south along Bimerick Creek are the roaring Bimerick Falls.

Between mileage markers 113 and 116 on Highway 12 are four waterfalls that are near the road and convenient to visit. Tumble Falls (to the west) drops 30 feet in a broad sheet before flowing beneath the highway. Horsetail Falls (to the east) drops nearly 100 feet into the Lochsa River. A mile beyond is spectacular Shoestring Falls which plummets for 200 feet in five stair-step plunges. Finally, Wild Horse Creek Falls, just before milepost 116, is a pair of falls that cascade about 50 feet side by side.

Another of the forest's special attractions is not far away. It is the 1.3-million-acre Selway-Bitterroot Wilderness. For many miles, Highway 12 is no more than five miles north, and from milepost 113 to 121 it forms one of the boundaries of the wilderness. Four different national forests share the wilderness. Hidden Lake and Big Sand Lake are two mountain lakes in the Clearwater section of the Selway-Bitterroot that can be reached from the trail head at Elk Summit. The historic Lochsa Ranger Station, in operation since the 1920s, is located along U.S. Highway 12 just north of the wilderness. It was restored in 1976 so that it appears the way it did a half century ago. It is open to the public during the summer months. Another historic forest building is the one-and-a-half-story log cabin and assorted outbuildings at the Weitas Guard Station. Clearwater Forest work crews use this remote cabin today. If you hike Forest Trail #20 south from the Weitas Creek Campground, you will come to these historical structures.

The Jerry Johnson Campground near the Lochsa River along Highway 12 is a serene place to make camp. You can count on the wind that whistles through the conifers to put you to sleep at night. Near the campground a trail runs along Warm Spring Creek to Jerry Johnson Hot Springs, one of several natural springs in the forest that emit very warm water. If you continue on the trail another two miles, you will come to Jerry Johnson Falls which drop 70 feet into a catch pool. The pounding of the falls deafens your ears long before you see it.

The DeVoto Grove of giant western cedars, also adjacent to Highway 12, deserves a visit. Huge cedars, some several hundred years old, grow above a densely shaded forest floor of ferns and wildflowers, giving one the feeling of being in a tropical rain forest. Another grove of giant cedars, the Heritage Grove, occurs along the Isabella River in the northwestern corner of the forest.

Highway 12 eventually climbs over the Bitterroot Mountains at Lolo Pass, the boundary between the Clearwater and Lolo national forests. A visitor's information center is maintained at the pass during the summer. From the pass you can walk or drive to Packer Meadows, a prominent stopping place during the Lewis and Clark expedition and an area where mountain wildflowers still bloom profusely today.

A final star attraction of the forest is the Mallard-Larkins Pioneer Area, a 30,500-acre region that spans the Clearwater and St. Joe national forests. It is a near-alpine area highlighted by sparkling, jewellike lakes that reflect adjacent snowy mountain peaks. Since the area is without roads, it is necessary to hike everywhere. Black Mountain, in the Clearwater portion of the area, is the highest elevation at 7,077 feet.

Coeur d'Alene National Forest

Location: 723,000 acres in northern Idaho, east of Coeur d'Alene. Interstate 90 is the major access. Maps and brochures available at the district ranger stations in Coeur d'Alene and Wallace or by writing the Forest Supervisor, 1201 Ironwood Drive, Coeur d'Alene, ID 83814.

Facilities and Services: Campgrounds (some with camper disposal service); picnic areas with grills; boat ramps; winter sports facilities.

Special Attraction: Settlers Grove of Ancient Cedars.

Activities: Camping, picnicking, hiking, backpacking, hunting, fishing, boating, winter sports, nature study.

Wildlife: Black bear, elk, bobcat.

AT THE EASTERN EDGE of the Coeur d'Alene National Forest, along the West Fork of Eagle Creek, is an extensive stand of giant western red cedars known as the Settlers Grove of Ancient Cedars. It is a prime attraction of the forest. Many of these giants tower above a rich assemblage of ferns and wildflowers. Several of the cedars are more than 500 years old. A hike beneath them is a great restorative remedy.

History buffs will want to stop at the Fourth of July Summit on Interstate 90 and take a short hike through a dense forest of towering western white pines to the Mullan Tree. The Mullan Tree was a giant white pine that Captain John Mullan's party encountered as they blazed the first trail through the area in 1861. A member of Mullan's crew carved "July 4, 1861, M.R." on the tree. This tree stood until November 19, 1962, when it broke off 20 feet above the ground during a windstorm. The tall stump, which is protected today by an iron cage, still bears the blaze on it. There are several short trails leading from the Mullan Tree that are fun to explore.

Even better trails are the Coeur d'Alene River National Recreation Trail and the Independence Creek National Recreation Trail. The Coeur d'Alene River Trail parallels the river for several miles, crossing a number of rushing tributaries. The Independence Creek Trail, several miles longer, begins at Weber Saddle on the Coeur d'Alene–Kaniksu border, passes through rugged terrain, and circles north around Griffith Peak.

For an overnight stay in the forest, we recommend the Devil's Elbow Campground nestled beneath scented pines on a big bend of the Coeur d'Alene River. After spending a night by the river, you may wish to hike next day along Yellow Dog Creek for about four miles to Fern Falls, a sparkling cataract.

There are several outstanding vistas in the forest. One favorite is the panoramic view from the top of Mount Coeur d'Alene. From the 4,439-foot summit you can get a good idea of the vastness of Coeur d'Alene Lake which sprawls around three sides of the mountain. Another sterling view overlooking the city of Coeur d'Alene is from West Canfield Butte. The butte can be reached by a back road northeast of the city.

Kaniksu National Forest

Location: 1,615,000 acres in northern Idaho and adjacent sections of Montana and Washington. U.S. Highways 2, 10A, and 95 and State Routes 31 and 57 are the major roads. Maps and brochures available at the district ranger stations in Bonners Ferry, Priest River, and Sandpoint or by writing the Forest Supervisor, 1201 Ironwood Drive, Coeur d'Alene, ID 83814.

Facilities and Services: Campgrounds (some with camper disposal service); picnic areas with grills; boat ramps; swimming beaches.

Special Attractions: Cabinet Mountains Wilderness; Northwest Peak Scenic Area.

Activities: Camping, picnicking, hiking, backpacking, hunting, fishing, boating, swimming, horseback riding, nature study.

Wildlife: Endangered American caribou; black bear, moose, elk.

WE BEGAN OUR EXPLORATION of the Kanisku National Forest by driving north out of the community of Priest Lake on Route 57 to Kalispell Point on the west bank of Priest Lake. From our vantage point, we could see Kalispell Island a mile offshore. The island, about one square mile in size, has five campgrounds, all reached only by boat. The 20-mile-long lake, in places nearly 5 miles across, is excellent for fishing, swimming, or boating. From Kalispell Point we drove 2 miles west to Hanna Flats where a nature trail has been laid out through a forest of giant western white pines. Because of the water-saturated soil in which these pines grow, the small area escaped a fire several years ago that demolished all other trees in the vicinity. Although the huge western white pines are impressive, the small plants that live in the understory are sure to delight the wildflower lover. The four-parted, white-flowered dwarf cornel and two slender, six-inch-tall orchids were among the plants we saw flowering in late June. Beyond Hanna Flats, Route 57 becomes a gravel road as it follows turbulent Granite Creek, past lovely, lily-laden Huff Lake, to the Roosevelt Grove of Ancient Cedars near the foot of High Rock Mountain. A campground in this secluded spot is dwarfed beneath the giant western red cedars that are among the biggest you will see anywhere. A pleasant surprise greeted us when we hiked about 50 yards from the campground to Granite Creek. Around a sharp bend and thundering for all it was worth was Granite Falls, a 50-foot drop over piles of huge boulders.

Next morning we were on our way north again on progressively poorer roads through the Selkirk Mountains until we were less than two miles from Canada. By trail we hiked another mile north to Upper Priest Falls, a violent 125-foot drop of the Upper Priest River. The Selkirk Mountains, incidentally, are the only place in the United States where the endangered American caribou still lives. There are thought to be less than 20 of them left.

Much of the Kaniksu surrounds gigantic Lake Pend Oreille. From Bernard Overlook at the extreme southern end of the lake is a breathtaking overview of the lake, which in places is more than 1,100 feet deep. A few miles northeast of Lake Pend Oreille, via the Lightning Creek Road, is Char Falls which plummets 75 feet with ear-shattering force.

We made a point of stopping at the forest's special attractions. One is the Cabinet Mountains Wilderness, shared with the Kootenai National Forest. It occupies the eastern edge of the Kaniksu. This 95,000-acre backcountry is a perfect place to find solitude. We tried the mile-and-a-half trail to Dad Peak and enjoyed the vistas along the way. Another forest attraction is the Northwest Peak Scenic Area in the northeast corner of the Kaniksu. It is divided with the Kootenai and is discussed in that forest's entry.

Nezperce National Forest

Location: 2,218,333 acres in north-central Idaho. U.S. Highway 95 and State Route 14 are the principal roads. Maps and brochures available at the district ranger stations in Elk City, Grangeville, Kooskia, and White Bird or by writing the Forest Supervisor, 319 E. Main St., Grangeville, ID 83530.

Facilities and Services: Campgrounds (some with camper disposal service); picnic areas with grills; boat ramps.

Special Attractions: Hells Canyon National Recreation Area; River of No Return Wilderness; Selway-Bitterroot Wilderness; Gospel Hump Wilderness.

Activities: Camping, picnicking, hiking, backpacking, hunting, fishing, boating, rafting, swimming, nature study.

Wildlife: Black bear, moose, elk, bighorn sheep, mountain goat, bobcat, mountain lion.

A BROCHURE from the Nezperce National Forest warns that two-thirds of the forest has no roads in it. We discovered that this was no understatement on our visit to the Nezperce. Except for State Route 14 which runs from Grangeville to Elk City and a relatively small number of back roads, the forest has to be hiked, boated, or rafted. There is ample opportunity for all of these old-fashioned methods of transportation in this forest of deep canyons, high jagged peaks, and some of the wildest water in the country. One of the forest's drawing cards is Hells Canyon. It is formed by the Snake River and straddles the Nezperce and Payette national forests in Idaho and the Wallowa-Whitman National Forest in Oregon. It is rated as the deepest gorge in the United States. From He Devil Mountain in the Nezperce to the Snake River below is a drop of 7,000 feet. He Devil Mountain is one of the wickedly rugged Seven Devils Mountains that tower over the eastern side of Hells Canyon. A road from Riggins, usually not open until July, climbs into the exciting Seven Devils area. Heavens Gate, at 8,500 feet, provides a grand view of the area. Beyond the Gate are extraordinary high mountain campsites at Basin, Shelf, Sheep Creek, Lower Cannon, and Seven Devils lakes. Prominent in the landscape is spirelike Devil's Tooth which projects skyward like a giant molar. Beyond the campgrounds, by way of a steep three-mile trail, is Mirror Lake which reflects the nearby so-called Tower of Babel in its glistening water. Tower of Babel is a prominent rock formation that rises high in the air. Most of the lakes in this area are excellent for trout and bass fishing. If you move noiselessly and speak in hushed tones, you may see black bear, bighorn sheep, and mountain goats. We prefer the coolness of the heavily forested mountains to the hot, desertlike Hells Canyon, but it is worth the small discomfort of descending into the canyon just to look up and see the immensity of the chasm. One of the best ways to do this is via float trips or jet boats in the Snake River. You can drive to the landing where you can make arrangements by taking a side road off U.S. Route 95, although steep and rough, to Pittsburgh Landing.

Another boast of the Nezperce is the River of No Return Wilderness, a part of the untamed Salmon River, that gouges its way 425 miles from southern Idaho to Snake River. It was not until about 1862 that white man successfully made it down the River of No Return. Meriwether Lewis was convinced it would never be conquered! Only the most hardy and experienced rafters, however, should try it as it has Class V challenges (on a scale of VI), such as Gun Barrel Rapids and Salmon Falls. The river's gorge is second deepest in the United States. Another Nezperce boast is the massive Selway-Bitterroot Wilderness which covers more than 1.3 million acres, some that run over into the Bitterroot, Clearwater, and Lolo national forests. The wild and wonderful Bitterroot Mountains are part of the eastern boundary. Complementing the mountains is the gushing Selway River which cuts through the Nezperce a few miles west of Gardiner Peak and forms a sheer-cliffed gorge before leaving the wilderness at Selway Falls. These falls, which are adjacent to a pine-scented campground, cascade for 50 feet over a rocky precipice. Of the many worthy trails in the wilderness, we like the one that parallels sparkling Moose Creek; it passes through a forest of giant and ancient western cedars. Colorful wildflowers, some growing only six inches tall, carpet the forest floor and provide a marked contrast to the towering cedars. Numbered among the wildflowers are delicate members of the lily and orchid families. Be on the lookout for some of the large elk herds that live near the Selway River. The forest service estimates about 200 different kinds of birds spend at least some time each year in the wilderness.

Still another Nezperce attraction is the Gospel Hump Wilderness, a roadless area whose 206,000 acres are totally within the forest. Some of the popular mountains are Gospel Peak, Buffalo Hump, North Pole, Quartzite Butte, Oregon Butte, and Square Mountain, all rising between 8,000 and 9,000 feet. Some have adjacent clear-water lakes such as the Gospel Lakes, Moores Lake, Fawn Lake, and Quartzite Lake. As you hike from the lower elevations near the Salmon River to the mountain summits, you will gradually pass from dry grasslands to a ponderosa pine forest to a Douglas-fir-dominated woods. At higher elevations the Douglas fir is replaced by lodgepole pine which, in turn, is replaced by alpine fir and whitebark pine. A myriad of mountain wildflowers is found both in the coniferous forests and in the many meadows. One fascinating plant that you will want to pause and examine is bear grass, whose unusual dense white plumes are borne on a stout, leafless stalk that rises three or four feet above a cluster of stiff, narrow leaves found at the base of the plant.

History buffs can follow the long and tortuous gravel road to the site of the once-prosperous mining community of Florence. All that is left is the solemn cemetery. It is hard to believe as you stand in this stillness that this was once a region of bustling commercial activity.

Payette National Forest

Location: 2,314,436 acres in west-central Idaho. U.S. Highway 95 and State Routes 14 and 55 are the access roads. Maps and brochures available at the district ranger stations in Council, New Meadows, McCall, and Weiser or by writing the Forest Supervisor, Forest Service Building, McCall, ID 83638.

Facilities and Services: Campgrounds (some with camper disposal service); picnic areas with grills; boat ramps; swimming beaches.

Special Attractions: Hells Canyon; Seven Devils Scenic Area; River of No Return Wilderness.

Activities: Camping, picnicking, hunting, fishing, hiking, backpacking, horseback riding, nature study, winter sports.

Wildlife: Endangered or threatened bald eagle, American peregrine falcon, and northern Rocky Mountain wolf; bobcat, Rocky Mountain bighorn sheep, mountain goat, black bear, wolverine, moose.

FOR THOUSANDS OF YEARS the Snake River has been cutting through brown basalt mountains and forming Hells Canyon, America's deepest gorge. The canyon, one of the drawing cards of the Payette, can be seen and explored many ways. One option is to drive the 26 miles up the canyon from Oxbow Dam to Hells Canyon Dam; the road stays in the canyon most of the way and offers many superb views. The day we took this road a young bald eagle swooped so low it almost hit our car.

Another option—only for the hardy—is to hike the six-mile trail from Hells Canyon to Kinney Point, one of the cliff summits, where there is a breathtaking panorama of the canyon. Another way to view the canyon is to drive the rough, narrow road to Sheep Rock, where there is also a short nature trail. A guide booklet at the trail entrance will help you identify the Rocky Mountain maple, western snowberry, subalpine fir, Douglas fir, whitebark pine, penstemon, and other plants that grow along the trail. Both of these viewpoints are in the Seven Devils Scenic Area, a rugged terrain above Hells Canyon that extends into the Nezperce National Forest.

For those who get their thrills driving hairy-scary back roads, try the Kleinschmidt Grade which connects the Hells Canyon Road with the diminished town of Cuprum by means of the narrowest and steepest road imaginable. If you are not used to mountain roads, do not try this one.

There are several areas out of McCall that are fun to explore. Brundage Mountain and Reservoir, the Hazard Lakes, and Hershey Point are northwest of McCall. To the southeast toward Burgdorf are trails and a road along the North Fork of the Payette River. Southward are additional scenic routes to the wild Salmon River which forms the northern boundary of the Payette. Most of the Salmon River, by the way, can be seen only by floating, but experienced outfitters should guide you because of the roughness of the water. East of McCall is the South Fork of the Salmon River. There is a good hiking or driving route that follows the river before branching north to Thunder Mountain and its neighboring Roosevelt Lake.

One secluded area we like is Jungle Creek in the West Mountains west of Cascade Reservoir. Jungle Creek, just over the pass to the west, is a fast-flowing stream that rushes through gorgeous dense vegetation. In the spring giant white and pink trilliums line the creek. Later the oranges and yellows of summer wildflowers abound.

More than 600,000 acres of the Payette are in the River of No Return Wilderness, one of the forest's prize attractions. Although the wilderness area is extremely rugged, dozens of trails crisscross the region. One trail that will take you to the heart of the wilderness proceeds east from the Big Creek Campground.

Mule deer, white-tail deer, elk, foxes, black bears, bobcats, and mountain lions are dominant wildlife in the Payette. Rare animals include Rocky Mountain bighorn sheep, mountain goats, wolverines, and moose.

Salmon National Forest

Location: 1,770,000 acres in east-central Idaho, around Salmon. U.S. Highway 93 and State Routes 28 and 29 are the major highways. Maps and brochures available at the district ranger stations in Leadore, North Fork, and Salmon or by writing the Forest Supervisor, Forest Service Building, Salmon, ID 83467.

Facilities and Services: Campgrounds (some with camper disposal service); picnic areas with grills; boat ramps.

Special Attraction: River of No Return Wilderness.

Activities: Camping, picnicking, hiking, backpacking, hunting, fishing, boating, swimming, horseback riding, winter sports.

Wildlife: Mountain goat, Rocky Mountain bighorn sheep, black bear, mountain lion, bobcat, elk, moose.

WE SUGGEST approaching the Salmon through its northern entrance on U.S. Highway 93 via Lost Trail Pass. A visitor's information map at the pass provides directions for proceeding into the forest. The highway to North Fork at times follows the North Fork of the Salmon River. Several trails and back roads lead off into the forest from Highway 93; the one to Granite Mountain is particularly scenic.

Most of the activities in the Salmon center about the wild and wonderful Salmon River and its turbulent tributaries. For 153 miles, from North Fork in the forest to Riggins in the Nezperce National Forest, the river thunders through steep-walled canyons as it rushes feverishly toward the Snake River. In places the canyon is deeper than the Grand Canyon, and ranks only behind the Snake River Canyon as the deepest in the country. Mountain goats frequently dot the cliffs that tower over the river. Shoshone Indians convinced George Rogers Clark in 1805 that the river was impassable; it took many years before the first river rafter made it safely through. Today campgrounds and hiking trails permit exploration at either end of the river, but the sheer cliffs in the central section can only be seen by rafting or floating. Because of violent rapids, drops of as much as eight feet, and other impediments, floating should be done only with some of the licensed outfitters in the area.

Exploring the Salmon River by foot or by car is recommended from North Fork to the Pine Rapids Bridge just beyond the nearly abandoned mining community of Shoup. Dilapidated wooden structures line the road and the river at

Shoup. (Clark, in his 1805 journey, turned back just before reaching the site where Shoup sprang up.) Several of the canyon walls are covered with Indian drawings. Beyond Shoup are Indian shelters that have been designated Historical Places. You may look at them and photograph them, but it is against the law to remove or deface anything in the area. There is a long, rough back road that heads north from the Salmon River road near Shoup. It eventually comes to Long Tom Mountain but offers an exceptional number of panoramic vistas of the forest.

As you might imagine, fishing for rainbow trout and Chinook salmon is excellent. Both fish spawn in the headwaters of the Salmon. The offspring migrate to the Pacific Ocean where they mature before returning to the stream where they were born.

The Middle Fork of the Salmon River bisects a portion of the River of No Return Wilderness, the big special attraction of the forest. We do not recommend taking the Middle Fork. It is a wild and unruly river and is passable only by floating. Only a few trails lead to the river, because of the extremely rough terrain of the region. As the river approaches the Salmon River, it passes through an area of large boulders known as Big Crags. One of the most scenic trails to the Middle Fork is from the Crags Campground in a remote setting at the edge of the wilderness. Because of the frequency of snow, the rough road to the campground usually is not open until July; snow often falls again by mid-August. The trail from the Crags circles 9,411-foot Cathedral Rock before winding past Welcome Lake, Heart Lake, and Terrace Lake on its way to the Middle Fork where there are several river campsites.

One campsite we particularly enjoyed is Meadow Lake Campground. It is set amidst dramatic mountain scenery in the southeastern corner of the forest, where Portland Mountain and Sheephorn Peak rise above 10,000 feet. The nights are refreshingly crisp at this 9,200-foot elevation, and a warm sleeping bag is a must.

Sawtooth National Forest

Location: 2,200,000 acres in south-central Idaho, extending into northern Utah. State Route 75 penetrates the northern unit of the forest, while Interstate 84 is the best access to the southern areas. Maps and brochures available at the district ranger stations in Burkey, Fairfield, Ketchum, and Twin Falls, at the Sawtooth National Recreation Area Visitor's Center, at the Redfish Lake Visitor's Center, or by writing the Forest Supervisor, 1525 Addison Avenue East, Twin Falls, ID 83301.

Facilities and Services: Campgrounds (some with camper disposal service); picnic areas with grills; boat ramps; swimming beaches; visitor's centers; lodge at Redfish Lake.

Special Attractions: Sawtooth National Recreation Area; Sawtooth Wilderness.

Activities: Camping, picnicking, hiking, backpacking, hunting, fishing, boating, swimming, horseback riding, nature study, winter sports.

Wildlife: Rocky Mountain bighorn sheep, elk.

WILD, SAWTOOTHLIKE MOUNTAINS dominate the northern part of the Sawtooth Forest between Ketchum and Stanley. We suggest a stop at the visitor's center north of Ketchum for an orientation for the Sawtooth. In the forest's recreation area there are a number of sights and experiences not to be missed. There are many hiking trails that lead to the Sawtooth Range, particularly in the vast Sawtooth Wilderness west of Stanley. The nearly four-mile trail to Bench Lake from Redfish Lodge climbs 1,000 feet to a gorgeous panorama of Redfish Lake. Trees at the higher elevations in the forest include whitebark and limber pines, with subalpine fir a little lower down. As the trails descend to 7,000 feet, Douglas fir, lodgepole pine, Englemann spruce, and ponderosa pine become prominent. Stands of quaking aspen punctuate the conifer forests. Black bears roam the forests and are sometimes seen. Bobcats, Canadian lynx, mountain lions, and the very rare wolverines are also here, but these last are seldom seen.

We also like the rugged Boulder Mountains north of the visitor's center. These mountains, which have layers of vividly colored rocks, reflect different colors as the sun passes overhead. Take the trail or jeep road to Boulder Basin. As you drop into the basin, you will be rewarded by the Boulder Lakes and the remnants of the old mining camp. This scene is enhanced during spring and summer by a colorful selection of wildflowers. There is another nice hiking trail from the Wood River Campground to the lovely Amber Lakes.

The scenic and turbulent South Fork of the Boise River twists east from nearby Featherville into the Sawtooth National Forest. Several campgrounds and trails are along the way. A rugged nine-mile trail leads up Big Water Gulch to Ross Peak and passes through some breathtaking scenery.

South of Interstate 84 are several disjunct units of the Sawtooth National Forest. The Albion Mountains south of Burley are dominated by 10,000-foot Cache Peak and the slightly shorter Mt. Harrison. There is a good access road into the Mt. Harrison area, but the only way you can reach Cache Peak is by hiking. Fourteen miles south of Hansen, the road enters the Cassia section of the forest and follows Rock Creek for many miles. From this road are several trails that lead into wild and woolly canyons, to bubbling springs, and to mountain ridges.

St. Joe National Forest

Location: 863,000 acres in northern Idaho, south and east of St. Maries. U.S. Highway 95A and State Routes 3 and 8 are the major roads. Maps and brochures available at the district ranger stations in Avery and St. Maries or by writing the Forest Supervisor, 1201 Ironwood Drive, Coeur d'Alene, ID 83814.

Facilities and Services: Campgrounds (some with camper disposal service); picnic areas with grills; winter sports facilities.

Special Attraction: Mallard-Larkins Pioneer Area.

Activities: Camping, picnicking, hiking, hunting, fishing, winter sports.

Wildlife: Black bear, elk, mountain goat.

ELEVEN MILES WEST OF AVERY, IDAHO, we turned onto Forest Road 312 from the St. Maries–Avery Road to begin a breathtaking journey up and over the Clearwater Mountains to Clarkia. After bouncing over a narrow corduroy-ribbed-like road for a few miles, the surface smoothed a bit, and we began to enjoy fully the swift, clear Marble Creek that paralleled our route. We stopped several times where marblelike rocks and other colorful stones drew our attention. The road goes through the heart of a heavily forested area that was sadly logged of nearly every western white pine during the 1930s. There are a number of remnants still to be seen of these ruthless early logging days. For instance, you will see the wooden pilings of several splash dams in Marble Creek that were used to transport logs downstream. The short trip to one site (marked on a forest map) is a mile and a half through a dense forest. Nearby is the rusted shell of a steam donkey, a threshing-machine-like engine used half a century ago in the logging operations. It is hard not to become nostalgic as you examine the splash dam and steam donkey. You may photograph these cultural artifacts if you wish, but forest laws protect them from molestation. At this point, in lieu of staying on the dirt road toward Hobo Pass, take a side road (marked on a forest map) that leads into a marvelous stand of huge and ancient western red cedars. It is known as the Hobo Grove Botanical Area. There has been little development here, and you have to blaze your own trail among the giants, but it is fun and well worth the effort. Finally, at Clarkia, Forest Road 312 joins State Route 3. If you have time, stay on the paved state highway to Bovill and take State Route 8 to Elk River. As you follow Elk Creek southward, you will encounter a startling series of six waterfalls within a distance of a half mile. The best and largest of the falls are Elk Falls and Lower Elk Falls. Elk Falls drops 150 feet from the canyon rim; Lower Elk Falls is 50 feet shorter, but gushes with greater force. At both falls, exercise great caution if you are observing them from the canyon rim, because it is easy to lose your footing.

The forest lists the Mallard-Larkins Pioneer Area 20 miles south of Avery as its prime attraction. It is a 6,400-acre roadless area that extends into the Clearwater National Forest. There are pristine glacial lakes and streams so pure that you see bottom. Unfortunately there are not many trails, but one good one is from the road's end at Buzzard Roost to Larkins Lake, one of the prettiest lakes you will see in a lifetime. In addition to its beauty, it contains challenging rainbow and cutthroat trout for anglers. It is possible to spot a black bear or a mountain goat in this unspoiled region.

Targhee National Forest

Location: 1,642,000 acres in eastern Idaho, west of Yellowstone and Grand Teton National Park; also a small segment in Wyoming. U.S. Highways 20, 26, 91, and 191 and State Route 3 are the primary access roads. Maps and brochures available at the district ranger stations in Ashton, Driggs, Dubois, Idaho Falls, and Island Park or by writing the Forest Supervisor, 420 No. Bridge St., St. Anthony, ID 83445.

Facilities and Services: Campgrounds (some with camper disposal service); picnic areas with grills; boat ramps; swimming beaches.

Special Attraction: Centennial Wild Area.

Activities: Camping, picnicking, hiking, backpacking, hunting, fishing, boating, swimming, horseback riding.

Wildlife: Rare trumpeter swan; endangered or threatened grizzly bear, bald eagle, and American peregrine falcon; sandhill crane, golden eagle, elk, moose, mountain goat, Rocky Mountain bighorn sheep.

THE TARGHEE is grizzly bear country. The grizzly, an endangered species, lives in the Lionhead area at the north end of the forest, and in the Fall River region off the southwest corner of Yellowstone Park. Lionhead is a peak on the Continental Divide which also has a small herd of Rocky Mountain bighorn sheep. The principal trees here are lodgepole pine and Douglas fir. The Fall River area, a region of uplands dominated by lodgepole pine, has several rough and scenic gorges. Because of a number of shallow lakes and marshes, this area is one of the two places in the United States where the rare trumpeter swan winters. The bodies of water also attract the small remaining populations of the sandhill crane. Cave Falls Campground is a good place to pitch camp and start your explorations. It is located near the end of State Route 36 just below Yellowstone's southern boundary; it not only is in a marvelous setting but also is near trails to Loon Lake, Swan Lake, and Widget Lake. It is convenient for forays into Yellowstone, as well.

The rugged Centennial Mountains, a part of the Continental Divide, form the northern edge of the Targhee and include the Centennial Wild Area, one of the forest's major attractions. The four-mile Howard Creek Trail off of Howard Springs Road brings an unobstructed view of Centennial Peak.

The eastern area of the forest includes the back of the Teton Range, where there are 11 peaks with an elevation greater than 10,000 feet. This region has glacier-scoured ridges and U-shaped canyons and is dominated by subalpine fir, limber pine, and spruce trees. At these high elevations are mountain goats, Rocky Mountain bighorn sheep, elk, and moose. The trail that goes over Buck Mountain Pass from Alaska Basin into Grand Teton National Park is spectacular. Alaska Basin is a special alpine area with dwarf alpine wildflowers in every color of the rainbow. Another excellent trail follows Darby Creek to South Fork Darby Canyon as far as Wind Cave. Because of the fragile nature of the cave, large groups are not permitted entry.

The southern and southwestern borders of the forest are formed by the Snake River. One place has been dammed to form the huge Palisades Reservoir. There are campgrounds and boat ramps spaced along the eastern shore of the reservoir. A rewarding trail exits Palisades Campground and proceeds along Palisades Creek to the two Palisades Lakes. After admiringly drinking in the beauty of the upper lake, and perhaps fishing for a while, hike south along Waterfall Canyon to its tumultuous waterfall. Highways 26 and 89 follow the Grand Canyon of the Snake River, the deepest gorge in the forest.

The far western part of the Targhee, lying on either side of State Route 28, has some interesting features. The Birch Creek pictographs, faded Indian writings stained red, are found in two rock shelters along rugged cliffs at the mouth of Indian Head Canyon, a tributary in the Birch Creek Valley. West of Birch Creek are 4 of the 16 charcoal kilns built around 1885 to supply fuel to an ore smelter nearby. The kilns, all made from local bricks, are about 20 feet high and 21½ feet in diameter.

Elsewhere in the Targhee, Lower Mesa Falls on Henry's Fork and Big Springs east of Mack Inn are sights not to be missed. A short trail leads to the 65-foot falls. Big Springs, which forms the headwaters of Henry's Fork, has a constant temperature of 52 degrees and flows at the rate of 186 cubic feet per second. Henry's Fork, by the way, is the best trout stream around.

CENTRAL ROCKIES

The national forests in the central Rockies total 19. They are the Arapaho, Grand Mesa, Gunnison, Pike, Rio Grande, Roosevelt, Routt, San Isabel, San Juan, Uncompahgre, and White River in Colorado; Ashley, Dixie, Fishlake, Manti-LaSal, Uinta, and Wasatch-Cache in Utah; and the Humboldt and Toiyabe in Nevada.

Arapaho National Forest

Location: 1,025,000 acres in north-central Colorado, west of Denver. Interstate 70, U.S. Highways 6 and 40, and State Routes 9 and 125 are the major roads. Maps and brochures available at the district ranger stations in Idaho Springs and Hot Sulphur Springs, at the Idaho Springs Vis-

itor's Center, or by writing the Forest Supervisor, 301 S. Howes, Fort Collins, CO 80521.

Facilities and Services: Campgrounds (some with camper disposal service); picnic areas with grills; boat ramps; concessions; visitor's center.

Special Attractions: Mt. Goliath Natural Area; Mt. Evans; Arapaho National Recreation Area; Indian Peaks Wilderness; Eagles Nest Wilderness.

Activities: Camping, picnicking, hiking, backpacking, hunting, fishing, boating, nature study, horseback riding, winter sports.

Wildlife: Elk, black bear, bighorn sheep.

AT IDAHO SPRINGS in the Arapaho is a small visitor's center with exhibits and a selection of literature about the forest. One suggested trip is the drive from the visitor's center past the nearly abandoned mining town of Alice to the end of the road; then you can hike from there about a mile uphill to St. Mary's Glacier and see its beautiful reflecting lake which is formed from the meltwater of the glacier. Your cares will melt away faster than the glacier at this solitary and tranquil place.

One of the forest's major attractions is the Mt. Goliath Natural Area. There is a trail at an elevation of between 11,500 and 12,000 feet that passes through a subalpine plant zone, finally reaching alpine vegetation. At the timberline are several specimens of bristlecone pine whose gnarled shapes are grotesque to the eye. In the alpine tundra are many bright-colored summer wildflowers. If you continue on from the rocky slopes of Mt. Goliath, you will come to another highlight of the Arapaho. It is 14,264-foot Mt. Evans, the highest mountain in the state of Colorado to have a road to its summit. Its summit is a most worthwhile excursion and can be reached by driving the paved road (State Route 5) or by hiking. Above the timberline are alpine meadows with gorgeous red, blue, yellow, and white low-growing wildflowers. We saw bighorn sheep, one with a most elegant set of curved horns, on our last visit.

Another popular attraction in the Arapaho is its National

Recreation Area with five major lakes, including Lakes Granby and Shadow Mountain. Boating is permitted on most of the lakes, and there are pleasant campgrounds throughout. We recommend Stillwater Campground because it has an amphitheater where naturalists give evening programs about the plants, animals, and geology of the area during the summer. The campgrounds also provide a superb view of the Indian Peaks Wilderness, another drawing card of the forest, where there are large areas of alpine tundra along with lakes and a few patches of glacier. The wilderness is divided between the Roosevelt and Arapaho national forests. Monarch Lake serves as the best trail head for trips into this wilderness from the Arapaho side. The Eagles Nest Wilderness, still another forest attraction, is a region of rugged, sharp ridges. It is shared with the White River National Forest and is discussed in that entry.

There are several beautiful mountain passes in the Arapaho. Two of them, Berthoud Pass and Loveland Pass, are on major U.S. highways. Along the south descent from Berthoud Pass is the Floral Park Picnic Ground which offers a sweeping view while you picnic. At Loveland Pass is Pass Lake Picnic Ground, one of the highest picnic areas in the United States, at 11,900 feet. But we prefer Hoosier, Boreas, and Guanella passes because of their unusual plant and animal life and abandoned mines.

Hoosier Pass, on State Route 9 south of Breckenridge, is where the Continental Divide runs east and west. A few hundred feet above the pass are moist alpine bogs that are spongy underfoot and contain several seldom-seen plants, including a tiny white-flowered member of the mustard family known as the Colorado eutrema, a plant being considered for the federal endangered species list. Boreas Pass is located on a scenic gravel road that connects Breckenridge with Como in the Pike National Forest. There are signs of abandoned mines everywhere, and near the Bakers Tank Picnic Ground there is a well-preserved tank that supplied water to ore trains until 1937. Guanella Pass is about nine miles south of Georgetown and is visited by bird-watchers; the pass serves as a wintering area for a unique bird known as the ptarmigan which is white in winter and dark in summer.

Grand Mesa National Forest

Location: 346,000 acres in west-central Colorado, east of Grand Junction and north of Delta. State Route 65 bisects the forest. Maps and brochures available at the district ranger stations in Collbran and Grand Junction, at the Grand Mesa Visitor's Center, or by writing the Forest Supervisor, 2250 Hiway 50, Delta, CO 81416.

Facilities and Services: Campgrounds (most with camper disposal service); picnic areas with grills; visitor's center; boat ramps.

Special Attraction: Grand Mesa, one of the world's largest flattop mountains.

Activities: Camping, picnicking, hiking, hunting, fishing, horseback riding, winter sports.

Wildlife: Elk, black bear, marmot, golden-mantled ground squirrel.

THE GRAND MESA NATIONAL FOREST is well named, because almost all of the forest is located on the Grand Mesa, one of the largest flattopped mountains in the world. Glacial activity on top of the mesa has scoured out bowl-shaped depressions that have filled with water, forming some of the best trout lakes anywhere. More than 300 of these alpine lakes are on the mesa. Many of them have lovely white water lilies on the surface. One expects to see a sprite emerge from the depths. Others have a fringe of majestic deep green conifers that contrast with the deep blue color of the glacial lakes.

Most of the mesa is between 10,000 and 11,000 feet. Spruces and firs predominate, but aspens make themselves known in the autumn when the mesa appears to turn a golden color. There are also scores of colorful wildflowers, including blue columbines, scarlet Indian paintbrushes, and various shades of penstemons.

We suggest that you start your exploration at the Grand Mesa Visitor's Center near the junction of State Route 65 and the road to Carp Lake. For an overview of the forest, follow Route 65 to Land's End. You can also see much of western Colorado from this overlook, which drops vertically 500 feet.

There are several excellent hiking trails on the mesa. Land O' Lakes Trail is a good one to warm up with since it is blacktopped and has easy grades during its half-mile loop. There are views of the mesa and many of the lakes from this trail. You may also glimpse marmots, chipmunks, and the golden-mantled ground squirrel.

Once adjusted to the high elevation, we recommend hiking the Crag Crest Trail which follows a picturesque ridge, or hogback, that is nearly 1,000 feet above the rest of the mesa. The trail is seven miles and at Eggleston Lake gradually climbs to 11,189 feet, the high point along the crest. Mountain meadows and thick spruce-fir forests perfume the air along the way.

For the experienced hiker or horseback rider, we recommend the 12-mile Kannah Creek Trail which zigzags from 6,000 to 9,000 feet up the western face of the mesa, through pinyon pine and juniper to oak and aspen, and finally to a stately stand of spruce and fir at Carson Lake.

There are several campgrounds in the forest, almost all of them along the shores of the alpine lakes and all of them above 9,000 feet. Sleeping out in the crisp air at this elevation insures a refreshing start next morning.

Gunnison National Forest

Location: 1.7 million acres in southwestern Colorado. U.S. Highway 50 and State Routes 114, 133, and 149 serve the forest. Maps and brochures available at the district ranger stations in Gunnison and Paonia or by writing the Forest Supervisor, 2250 Hiway 50, Delta, CO 81416.

Facilities and Services: Campgrounds (most with camper disposal service); picnic areas with grills; boat ramps.

Special Attractions: Gothic Natural Area; Slumgullion Earth Flow; West Elk Wilderness; La Garita Wilderness.

Activities: Camping, picnicking, hiking, backpacking, hunting, fishing, boating, nature study, horseback riding, winter sports.

Wildlife: Mountain sheep, elk, black bear.

GUNNISON NATIONAL FOREST, on the western edge of the main range of the Rocky Mountains, is a mecca of superior natural areas and spectacular scenery. There are also ghost towns and abandoned mines to add to the exploration of the forest.

The Gothic Natural Area is one of the places of special attractions and is located at the northern edge of the forest about three miles north of the tiny community of Gothic. This natural area contains a prime example of a virgin stand of Engelmann spruce and subalpine fir that can be seen at elevations ranging from 10,000 feet along the East River to 12,809 feet on the summit of Mt. Baldy. In addition to the pristine biological features, there are several alpine lakes and glacier-formed valleys. The Gothic Campground nearby is a good place to pitch camp and from which to explore the natural area. It is also adjacent to the head of Copper Creek Trail, a super-scenic route that passes cascading Judd Falls, sparkling White Rock Mountain, and serene Copper Lake before crossing East Maroon Pass. Beyond this is the Maroon Bells–Snowmass Wilderness of the White River National Forest.

We also heartily recommend the Slumgullion Campground along State Route 149 in the southern unit of the Gunnison. You get a feeling of exhilaration after sleeping in the thin, pure air at a high elevation such as this. This campground, at 11,500 feet, is among the highest in the country. Once there, hike west along Highway 149 to Windy Point Overlook and witness to the north a geological phenomenon known as the Slumgullion Earth Flow. About 700 years ago volcanic material on the mountain ridge, altered by gases and superheated water, began to flow as a thick liquid into the valley below. Some of this earth flow eventually built a natural dam that formed Lake San Cristobal a few miles west and out of the forest. You can still see this massive earth flow area that is moving at a rate of about 20 feet per year. It can be detected by the trees that lean in various directions.

There are two other trips we suggest in the Gunnison. Both can be driven or hiked. One leaves from the Erickson Springs Campground and proceeds for many miles to the community of Crested Butte. Near the beginning of this journey, Marcellina Mountain, a light-tan-colored mountain, towers magnificently above. The route parallels clear mountain streams and blue lakes and skirts massive mountains, sometimes winding through forests, other times crossing wildflower-laden meadows. Two-thirds of the way to Crested Butte is Kebler Pass, adjacent to an old cemetery claiming the remains of many pioneers. A one-mile side trip south is Ohio Pass where we saw an exceptionally large, splendid colony of blue columbine.

The other excursion starts a few miles east of Crested Butte near Jack's Cabin and follows Taylor Canyon as far as Taylor Park Reservoir, a large lake with sweeping views of the Matchless Mountains to the west. River's End and Lakeview campgrounds are at the north and south ends of the trail's reservoir. The morning we were there, we woke to a ghostly haze that filled the entire valley. Beyond the lake and through the more open Taylor Park, there is a long and tortuous climb to Cottonwood Pass at 12,126 feet. Just before the pass, the Timberline Overlook offers unparalleled views to the west. From a parking lot at the pass, however, a steep trail ascends to the mountaintop 100 yards away. This is one of the best vistas we have ever seen in the West. One hates to leave it, because it is a view one could savor almost forever.

Another side trip from the Taylor Park Reservoir is 10 miles south to the remains of the once-thriving mining camp of Tincup. From Tincup both road and trail go south, following Willow Creek and past Annie and Porcupine gulches. There is a sharp climb to Cumberland Pass, but there are startling views. Descending from this pass, the route parallels North Quartz Creek to the Quartz and Pitkin campgrounds before reaching Pitkin. East from the Pitkin Campground along Middle Quartz Creek for about 12 miles where Forest Road 839 disappears is the Old Alpine Tunnel. This tunnel, at 11,524 feet, was built at a cost of $242,000 in 1880, and is the highest and most expensive narrow-gauge railroad tunnel ever constructed.

A further special attraction of the Gunnison is the West Elk Wilderness, 62,000 acres of wild and wonderful terrain. It is a region of high mountain meadows, deep canyons, and densely forested slopes. Although West Elk Peak reaches an impressive 13,035 feet, the sheer-walled towers of the rock formations known as "The Castles" nearby are the most spectacular sight in the wilderness. The Castles are best viewed from the Mill Creek Trail which climbs above timberline.

One further attraction of the forest not to be missed is the La Garita Wilderness which is shared with the Rio Grande National Forest. This is a high mountain wilderness dominated by San Luis and Stewart peaks, both at about 14,000 feet.

Pike National Forest

Location: 1,105,000 acres in central Colorado, west of Colorado Springs. U.S. Highways 24 and 285 and State Routes 9, 67, and 126 are the major highways. Maps and brochures available at the district ranger stations in Colorado Springs, Fairplay, and Lakewood or by writing the Forest Supervisor, 1920 Valley Drive, Pueblo, CO 81008.

Facilities and Services: Campgrounds (most with camper disposal service); picnic areas with grills.

Special Attractions: Pikes Peak; Lost Creek Scenic Area; Lake Abyss Scenic Area; Windy Ridge Bristlecone Pine Scenic Area.

Activities: Camping, picnicking, hiking, backpacking, hunting, fishing, boating, horseback riding, nature study, winter sports.

Wildlife: Rocky Mountain bighorn sheep, elk, black bear, antelope.

MOST VISITORS to the Pike National Forest have well-known Pikes Peak at the top of their list. Although there are many higher peaks in the United States, it is an outstanding landmark of the Great Plains region. The summit of this nationally famous 14,110-foot mountain can be reached by car, by cog railroad, or on foot. In addition to the fame of its height, visitors will discover an extensive alpine tundra above timberline where low-growing multicolored wildflowers and dwarf grasses prevail. As one approaches the summit, this zone gives way to a bleak and harsh one of rocks and patches of snow. At the Summit House there are concessions, as well as information for visitors. Hikers to the summit should use the Barr National Recreation Trail from Manitou Springs. It is an 11½-mile climb to the top.

After a visit to Pikes Peak, there are a number of options. If you go to the Colorado Springs area, there is an exciting and sometimes scary trip on Gold Camp Road to the once-booming mining towns of Victor and Cripple Creek. This road follows the route of the railway that transported gold from the mountains to the city. Even at the starting point, where the North Cheyenne Canyon Road intersects the Gold Camp Road, there is a sweeping view of the noisy Silver Cascades which drop over a sheer rock face penetrated by an old railroad tunnel. The Gold Camp Road continues through tunnels, up switchbacks, past abandoned towns and mines, always with spectacular scenery all around. After about 25 miles, the route leaves the Pike National Forest, but it continues on to the towns of Victor and Cripple Creek.

Another route that can raise your hackles is from Colorado Springs via the Rampart Range Road to the Rampart Reservoir Recreation Area. The road is rough and narrow in spots, but the breathtaking vistas along the way make the trip worthwhile. At the Rampart Reservoir are a boat ramp, several congenial picnic grounds, and a 13-mile Lake Shore Trail that encircles the reservoir.

Although the huge Eleven-mile Reservoir west of Florissant is nearly all on state land, the Pike National Forest does maintain campgrounds and picnic areas at the southeast corner of the lake as well as all along the Eleven-mile Canyon Road that leads to it. About 20 miles north is Lost Creek Scenic Area, a must on your hiking itinerary. Located in the rugged Tarryall Mountain Range, this 15,000-acre roadless area appears to have an endless supply of bare and sometimes amusing granite formations in every imaginable shape. Bighorn sheep are often seen climbing over the domes, spires, and pinnacles. Lost Creek itself disappears several times into granite sinks, only to reappear downstream. Closest access to the scenic area is from the Goose Creek Campground on Forest Road 211.

Another forest attraction is Abyss Lake at the northernmost extension, on a high western slope of Mt. Evans. The lake, which is 12,550 feet above sea level, lies in a deep rock-rimmed gorge, with Mt. Evans in the background. The best approach by trail is from the road to Mt. Evans, but this requires a long, roundabout drive through the Arapaho National Forest. To reach Abyss Lake from the Pike, follow the long trail up Scott Gomer Creek and Lake Fork from the Burning Bear Campground.

A further Pike attraction is in another direction—in fact about as far west as you can go. It is the Windy Ridge Bristlecone Pine Scenic Area on the blustery slopes of giant Mt. Bloss. In an open, parklike setting, twisted-trunked bristlecone pines dot the landscape. Bristlecone pines live to be the oldest organisms in the world, and their gnarled and twisted trunks even make the younger ones look old. On the way to Windy Ridge, between Forest Road 416 and Buckskin Creek, is a relic of interest from history. It is an arrestra, a stone used by the Spanish settlers in the 17th century to crush ore. This arrestra dates back to 1680.

Rio Grande National Forest

Location: 1,800,000 acres in southwestern Colorado, mostly west of Monte Vista, Del Norte, and Alamosa. U.S. Highways 160 and 285 and State Route 149 are the major access routes. Maps and brochures available at the district ranger stations in LaJara, Creede, Del Norte, and Saguache or by writing the Forest Supervisor, 1803 West Highway 160, Monte Vista, CO 81144.

Facilities and Services: Campgrounds (some with trailer disposal service); picnic areas with grills; boat ramps; archery range at Rock Creek.

Special Attractions: Wheeler Geological Area; La Garita Wilderness; Weminuche Wilderness.

Activities: Camping, picnicking, hiking, backpacking, hunting, fishing, boating, horseback riding, archery.

Wildlife: Elk, bear, bobcat, ptarmigan, blue grouse.

RIO GRANDE boasts of several unusual rock formations. The most spectacular are at the Wheeler Geological Area just outside the La Garita Wilderness and about 10 miles northeast of Creede. Here are a collection of strange shapes christened with such names as Temple, Milk Bottles, Beehives, Cyclops, Dante's Lost Souls, Pipe Organ, and Finger

Rocks. These unusual rocks are extremely fragile and it is important that they not be climbed or disturbed. In fact, the forest service has posted signs forbidding this, and rangers frequently patrol the area. The formations are made of coarse volcanic material called tuff that was spewn from an erupting volcano aeons ago. A jeep trail from Wagon Wheel Gap on Route 149 via Hansen Mills comes near the area, and there are several hiking trails including a 15-miler from Creede, past Inspiration Point, Mammoth Mountain, and Wason Park. Wheeler is also reached by a trail from the La Garita Wilderness.

Another special feature of the forest is the 49,000-acre La Garita Wilderness which is shared with the adjacent Gunnison National Forest. This ruggedly beautiful area of alpine meadows, rushing streams, and high peaks, including San Luis and Stewart peaks, each more than 14,000 feet, is a perfect place for solitude. Our favorite area is Machin Lake, nestled away from the maddening crowd at the foot of a mountain. It is a rewarding fishing spot for cutthroat and brown trout, as are most of the mountain streams in the wilderness. The ptarmigan and blue grouse are two birds indigenous to this high country. The best access to La Garita is from the Stonecellar Campground.

Both the La Garita and another major attraction, the Weminuche Wilderness, straddle the Continental Divide, the imaginary line that divides the water flow in the continent of North America. The Weminuche Wilderness, about half of which is in the San Juan National Forest, is much larger than the La Garita. Alpine lakes on the Rio Grande Forest side of the wilderness, such as the Ute Lakes, Twin Lakes, and Squaw Lake, are alive with trout. A good base for hiking into the wilderness is the Thirtymile Campground at the east end of Rio Grande Reservoir near Forest Road 520. The trail follows Weminuche Creek before crossing into the San Juan National Forest where there are good views of the Rio Grande Pyramid, a conspicuous pointed rock. Another scenic wonder is the Window, a gigantic opening that resembles a window in a granite cliff near the border into the San Juan.

There is much to see and do outside the wilderness areas. If you like falls, there are several along Clear Creek east of State Route 149 above Spring Creek Reservoir. You will be able to camp nearby as there is a choice of three facilities, all in splendid settings in the immediate vicinity.

Another suggested stopping-off place is Alamosa Canyon, a drive or long-distance hike west to Elwood Pass on the Continental Divide. Alamosa River, a yellowish stream that has no fish because of the high mineral content of the water, sculptured out Alamosa Canyon. Several abandoned gold mines, such as the Rosie S., Lulu, and I-Don't-Care, are still evident in the canyon. After about 10 miles, the canyon is interrupted by the wild, rushing Wightman Fork stream which enters the Alamosa River from the north. For a side trip, hike along this mountain stream for 5 miles to Summitville, site of the highest mining camp in Colorado.

If you continue along the main route, you will note the highly mineralized yellow waters of Bitter Creek which empties into the Alamosa just before reaching the Stunner Campground. The campground is surrounded by great numbers of purple gentians and other wildflowers and thus is a particularly soothing spot to make camp. Beyond Stunner the route swings out of Alamosa Canyon and provides a view to the north of Lookout Mountain, a lofty mass whose colors range from red to yellow, blue to black, and sometimes purple, depending upon the angle of the sun. Due west, on the Continental Divide, are the treeless summits of Montezuma Peak and Summit Peak, both more than 13,000 feet. Hikers will enjoy following Iron Creek to Elwood Pass on the Continental Divide, while those in vehicles should continue on Forest Road 380 to the pass. At Elwood Pass hike for another 10 minutes to the summit, through a fragrant spruce forest laden with wildflowers, ferns, and mushrooms. A few days in the area are a refreshing and relaxing experience.

Roosevelt National Forest

Location: 788,000 acres in north-central Colorado, west of Fort Collins and Boulder. Interstate 70, U.S. Highways 34 and 287, and State Route 14 are the major access roads. Maps and brochures available at the district ranger stations in Boulder and Fort Collins or by writing the Forest Supervisor, 301 S. Howes, Fort Collins, CO 80521.

Facilities and Services: Campgrounds (some with camper disposal service); picnic areas with grills; boat ramps.

Special Attractions: Rawah Wilderness; Indian Peaks Wilderness; Cache La Poudre Wilderness.

Activities: Camping, picnicking, hiking, backpacking, hunting, fishing, boating, nature study, horseback riding.

Wildlife: Elk, bighorn sheep, black bear.

ROOSEVELT NATIONAL FOREST is named after President Teddy Roosevelt, an ardent champion of outdoor life. It has four areas not to be missed. One is the Rawah Wilderness, which covers 27,000 acres in a remote northwest section of the forest. The wilderness is noted for its many high alpine lakes, clear mountain streams, and rugged peaks reaching nearly to 13,000 feet. The Chambers Lake Recreation Area outside the southeast corner of the wilderness is a good place from which to explore the Rawah. The relatively short Blue Lake Trail begins at the recreation area and goes to Chambers Lake, a postcard-pretty lake where the trout are known to bite. The lake is suitable for boating and has a pleasant and clean, though sometimes crowded, campground. Its beauty in a lonely setting sends gooseflesh over your body. One of the most spectacular sights in this wilderness is Iceberg Lake. For further solitude and a view of the Rawah Wilderness, camp at Free Water Bridge about 12 miles north of Chambers Lake on Forest Road 190.

Another worthwhile attraction of the Roosevelt National Forest is the Indian Peaks Wilderness across the southern boundary of Rocky Mountain National Park. Shared with the Arapaho National Forest, this wilderness' claim to fame is its Isabelle Glacier, a remnant of a once-great ice mass. There are also multitudinous beautiful, high mountain lakes that were formed when the glaciers retreated. Most visitors find the Brainerd Lake Recreation Area at the eastern edge of the wilderness a good base. It is only about five miles by trail from here to Isabelle Glacier by way of Long Lake and Lake Isabelle. There is much alpine tundra in the Indian Peaks Wilderness, where stunted wildflowers grow on the rocky surface.

The Cache la Poudre Wilderness is another special forest

attraction and is a wild, roadless area where solitude prevails. You can hike for miles and not see other life except for an occasional wild animal.

For a truly nostalgic as well as exciting journey, if you have the time, drive the route of the old Moffat Railroad from its eastern end at the Jumbo Mountain Rest Area on State Route 72 to the twin trestles at the Devil's Slide, some 18½ miles away. (The road used to go across the hackle-raising trestles and climb over Rollins Pass, but it is not safe for automobile travel anymore.) The original railroad was built in 1905 and continued in operation until the 6-mile Moffat Tunnel was bored through the Continental Divide. Along the old route you can still see abandoned buildings, an old water tower, and Yankee Doodle and Jenny lakes. There are scenic overlooks at every turn. When you see the twin trestles 1,000 feet above Middle Boulder Creek, picture yourself crossing them on a train. It makes your skin tingle and takes a few years off your life.

Routt National Forest

Location: 1,127,164 acres in northwestern Colorado, in the vicinity of Craig, Steamboat Springs, and Kremmling. U.S. Highway 40 is the major access route. Maps and brochures available at the district ranger stations in Craig, Kremmling, Steamboat Springs, Walden, and Yampa or by writing the Forest Supervisor, Hunt Building, Steamboat Springs, CO 80477.

Facilities and Services: Campgrounds (some with camper disposal service); picnic areas with grills; boat ramps.

Special Attractions: Mt. Zirkle Wilderness; Flat Tops Wilderness.

Activities: Camping, picnicking, hiking, backpacking, hunting, fishing, boating, horseback riding, nature study, cross-country skiing.

Wildlife: Bighorn sheep, elk, bobcat, black bear, mule deer.

IF IT IS HIGH COUNTRY you are after, you will find it in the Mt. Zirkle Wilderness in the Routt National Forest. Straddling the Continental Divide, the imaginary line that separates the flow of water to the Atlantic and Pacific, these 72,500 acres of wild country have 14 peaks in the 12,000-foot range, topped by Mt. Zirkle at 12,180 feet. There is some snow here the year around, and the entire wilderness is usually snow-covered until the end of June. The fragrant forests of spruce and fir give rise to colorful alpine meadows and rocky tundra above timberline. There are more than 65 sparkling lakes in the wilderness, all with brook, rainbow, and cutthroat trout, and a few with the more elusive California golden trout. Elk, mule deer, bobcat, and black bear are likely to be seen, and even the rare bighorn sheep with their characteristic curved horns are sometimes observed. When hiking in this wilderness, be sure to take proper precautions against the high altitudes by pacing yourself. The easiest access to the wilderness is from Slavonia near the western boundary. Seedhouse and Box Canyon campgrounds are convenient nearby. One of the trails from Slavonia to Mt. Zirkle passes Gilpin Lake, a refreshing body of water that is pleasant to contemplate.

The most popular attraction in the Routt is Rabbit Ears Peak, so named because of its twin rabbit-ear-like spires. It has served as a landmark to travelers for decades and is located southeast of the Mt. Zirkle Wilderness. A two-mile trail from Dumont Lake to the peak is rewarding. There are picnic areas and campgrounds around the lake, which is just north of U.S. Highway 40.

Southeast of Steamboat Springs is another forest drawing card. It is Flat Tops Wilderness, an area of high, flattopped plateaus through which flows the White River. The rolling tops of these plateaus alternate between grasslands and tracts of forest, but the canyons that gouge deeply into the plateaus form huge rocky amphitheaters. Much of this high plateau country is a part of the Flat Tops Wilderness, which, incidentally, is in both the Routt and White River national forests. The most unusual formation in the Routt section is the Devil's Causeway, a beguiling narrow rock ridge that has sheer drop-offs on either side. The causeway is reached by a rugged 15-mile trail (#1119) that originates from the Pyramid Guard Station on Forest Road 16 and follows the East Williams Fork until it is routed past Round Lake, Causeway Lake, and Little Causeway Lake. The Devil's

Causeway is on the ridgetop south of this trail near Little Causeway Lake. It can also be reached from the Cold Springs Campground at the edge of Stillwater Reservoir outside the wilderness boundary.

A geological feature of the forest is the great slide along Muddy Creek where silt is continuously being deposited into Muddy Creek. This phenomenon is visible to the west from the Morrison Creek Road (Forest Road 270) a few miles northwest of the Lynx Pass Campground, or it can be hiked to from Yampa by way of Eagle Rock Lakes and Wheeler Basin.

The extreme southeastern corner of the Routt is separated from Rocky Mountain National Park by the Continental Divide. The Divide passes through the Never Summer Mountain Range dominated by Mt. Cumulus, Mt. Cirrus, Mt. Richthofen, and Howard Mountain, all above 12,700 feet. True to its name, the range never enjoys a warm summer day. Trail #1141 from Lake Agnes to Baker Pass skirts the western slopes of these massive mountains.

Northeast of Steamboat Springs are the volcanic peaks of the Elkhorn Mountains. They are worth a trip if time permits and can be best reached from Craig by following State Route 13 and then Forest Road 110 to Lost Park. Sawmill Creek Campground between Black Mountain and Slide Mountain is along the way.

San Isabel National Forest

Location: 1,100,000 acres in south-central Colorado, from Leadville nearly to the New Mexico border. U.S. Highways 24 and 50 and State Routes 12, 62, 91, and 165 serve the forest. Maps and brochures available at the district ranger stations in Canon City, Leadville, and Salida, at the Monarch Pass Visitor's center or by writing the Forest Supervisor, 1920 Valley Drive, Pueblo, CO 81008.

Facilities and Services: Campgrounds (some with camper disposal service); picnic areas with grills; boat ramps; visitor's center.

Special Attractions: Mt. Elbert, highest peak in Colorado; about 20 peaks over 14,000 feet; Collegiate Peaks Wilderness.

Activities: Camping, picnicking, hiking, backpacking, hunting, fishing, boating, horseback riding, nature study.

Wildlife: Rocky Mountain bighorn sheep, mountain goat, elk, black bear, antelope, mountain lion, bobcat.

SAN ISABEL is a paradise for mountain climbers and has two outstanding offerings. They are Mt. Elbert, Colorado's highest at 14,431 feet, and Mt. Massive, only 13 feet shorter; each can be ascended and descended in a single but long day. Furthermore, it is possible to climb these two highest mountains in Colorado without ever having climbed a mountain before. You must be in good physical condition, however, to hike the required distance. You have a choice of two trails to the summit of Mt. Elbert, one from the north and the other from the south. The south approach, 7 miles long, is 2½ miles longer than the north trail, but is less difficult. On each trail timberline is passed at about 11,600 feet. From then on the trails lead through alpine meadows and over rocky surfaces. More gargantuan is Mt. Massive, which has seven peaks over 14,000 feet and more than 70,000 acres above timberline, but conquering it is no more difficult than conquering Mt. Elbert. Despite these two giants, we like the slightly lower Spanish Peaks which seem taller because they rise up abruptly from the surrounding plains. These two peaks have conspicuous ridges, called dikes, which radiate from the mountains and can be easily observed from the distance. Some of the dikes are as much as 100 feet tall, 100 feet wide, and up to 14 miles long. They are considered to be unusual formations. An overview of the Spanish Peaks is from the road or trail east of Apishapa Pass. The timberline on the West Peak can be reached by a 3-mile trail from the pass that winds through forests of bristlecone, limber, and ponderosa pines, alpine, white, and Douglas firs, and Engelmann spruce.

San Isabel's other special offering is the Collegiate Peaks Wilderness; it lies partly in the White River National Forest, but the Isabel side encompasses several of the impressive 14,000-foot peaks. There is a visitor's center at Monarch Pass where information about this wilderness and the forest in general can be obtained.

Venable Falls, a scenic cascade, and the spirelike, 14,191-foot Crestone Needle are two further places you might enjoy visiting. The Needle is only two miles via the South Colony Creek Trail from the end of a jeep road, or about nine miles from an ordinary road. In the vicinity are a number of limestone caves, but accessibility to them should be checked with the district ranger.

Many splendid lakes are in the forest, ranging from the large Turquoise and Isabel lakes with developed facilities to tiny, unnamed ones. Lake-in-the-Clouds, 10 miles west of Westcliffe, is particularly breathtaking and leaves your head in the clouds.

San Juan National Forest

Location: Nearly two million acres in southwestern Colorado, north of Durango and Cortez. U.S. Highways 84, 160, and 550 and State Route 145 are the main roads. Maps and brochures available at the district ranger stations in Durango, Dolores, Mancos, Pagosa Springs, and Bayfield or by writing the Forest Supervisor, 701 Camino del Rio, Durango, CO 81301.

Facilities and Services: Campgrounds (some with camper disposal service); picnic areas with grills; boat ramps.

Special Attractions: Weminuche Wilderness; South San Juan Wilderness; Lizard Head Wilderness; Chimney Rock Archeological Site.

Activities: Camping, picnicking, hiking, backpacking, hunting, fishing, horseback riding, boating, jeeping.

Wildlife: Endangered or threatened bald eagle and American peregrine falcon; elk, bighorn sheep, black bear.

A GOOD INTRODUCTION to the San Juan is provided by a ride on the Denver and Rio Grande Narrow Gauge Railroad which runs for 44 miles from Durango to Silverton and follows the Animas River as it passes through the San Juan National Forest. The trip affords visitors sweeping views of its star feature, the Weminuche Wilderness, with its West Needle Mountains, and several canyons with rushing mountain streams. Visitors to the Weminuche Wilderness may leave the train at Needleton Siding and hike 5 miles up Needle Creek to the Chicago Basin where the 14,000-foot peaks of Mt. Eolus, Sunlight, and Windom tower overhead. Or visitors can enter the Weminuche from the Vallecitos Reservoir. There are several campsites along the shore of this large lake that serve as good base camps. A trail from the Vallecitos Reservoir along the Los Pinos River enters the wilderness. From here the Lake Creek Trail goes to the dazzling blue Emerald Lake. A big attraction of the Weminuche is Window Rock, where a giant hole has been formed in a sheer rock wall and serves as a window along the Continental Divide.

Another forest attraction is the South San Juan Wilderness. At the extreme western edge of the San Juan, a few miles east of Dove Creek, is a picnic area that overlooks Dolores Canyon 1,500 feet below. It is a pleasant introduction to this wild region. Southeast of here, in a rough topography that includes impressive mesas, is an area known as Narraguinnep where some of the best examples of ponderosa pine, pinyon pine, western juniper, and oak brush can be seen. Aspens line the bottom of the canyon. Throughout the San Juan, at lower elevations, the vegetation is mostly ponderosa pine, pinyon pine, western juniper, Gambel oak, and sagebrush. In the cooler, higher regions, Douglas fir, white fir, cork-bark fir, Engelmann spruce, and quaking aspen dominate.

Still another attraction is Lizard Head Wilderness, a smaller wild area that is shared with the Uncompahgre National Forest. It features huge rocky outcrops, some with amusing shapes. It also has a score of high mountains, glacial lakes, and babbling mountain streams. There is also an abundance of alpine meadows and spruce-fir forests. Although Navajo Lake is a scenic attraction, it is poor for fishing.

There are several archeological sites in the forest where important Anasazi Indian ruins have been found. The best of these is at Chimney Rock, but you will need a forest service guide to take you there because of the fragile nature of the area. You also have a good chance of seeing the endangered American peregrine falcon here. Another site with prehistoric significance is along Spring Creek.

If you have a four-wheel-drive vehicle, one of the most breathtaking trips leaves U.S. Route 550 north of Silverton and climbs steeply over Ophir Pass to the small town of Telluride. The section of the road from Ophir Pass to Telluride is in the Uncompahgre National Forest.

Uncompahgre National Forest

Location: 944,000 acres in south-central Colorado, mostly south of Montrose. U.S. Highway 550 and State Route 145 are the major roads. Maps and brochures available from the district ranger stations in Montrose and Norwood or by writing the Forest Supervisor, 2250 Hiway 50, Delta, CO 81416.

Facilities and Services: Campgrounds (most with camper disposal service); picnic areas with grills.

Special Attractions: Mt. Sneffels Wilderness; Lizard Head Wilderness.

Activities: Camping, picnicking, hiking, backpacking, hunting, fishing, jeeping, horseback riding.

Wildlife: Mountain sheep, elk, black bear, mountain lion, lynx, bobcat.

THE UNCOMPAHGRE NATIONAL FOREST, on the western slope of the Continental Divide, is in two distinct units. Northwest from Ouray is the Uncompahgre Plateau, a rolling, high-elevation mesa occasionally penetrated by rugged canyons. Pinyon pine and western juniper dominate the lower areas, with aspen, spruce, and ponderosa pine in the higher elevations. South, west, and east from Ouray is the other unit at the northern edge of the San Juan Mountains. These mountains are jagged-peaked and intersected by deep narrow canyons; no wonder this magnificent section of the forest is often referred to as the "Switzerland of America." Most of the spectacular features of the forest are here.

Scenery is all-consuming in the Uncompahgre. One major attraction is the Mt. Sneffels Wilderness. Mt. Sneffels, the lordly 14,150-foot peak that towers above the shimmering Blue Lakes, is the focal point of the wilderness. The other big drawing card in the forest is Lizard Head Wilderness. Lizard Head, with its unusual rocky, spiraled peak, is the center of attraction in this wilderness, a region shared with the San Juan National Forest. Both of these mountains are reached by way of scenic trails.

Some mountains in the forest are highly colored, such as Red Mountain, which is conspicuous from Highway 550 a few miles south of Ouray. Then there is often-pink-looking Pinnacle Ridge which is sandwiched between the Middle and East forks of the Cimarron River. The trails that parallel each river fork provide continuous opportunity to observe this formation. This ridge is not as high as the surrounding mountains, so that trees usually grow nearly to its top, producing a striking color contrast with the rocks. Other mountains, like the incomparable Uncompahgre Peak, are not spotted by color but project prominently above their surroundings. This 14,309-foot monolith can be reached by a three-mile hike from the end of the Nellie Creek jeep road.

Since much of the area around Ouray and Telluride is mining country, hikers will come upon many abandoned mining structures in the mountains. While these old mines add a nostalgic touch of scenery, they can also be hazardous, so be careful if you decide to explore around any of them. You can fall into open pits and other dangers. Chief Ouray Mine in the cliffs above the town of Ouray offers superb views. The Cascade Trail leads to the mine from Amphitheater Campground east of town.

Many mountain lakes are in the Uncompahgre. Our favorite is Lake Hope, a small alpine fishing lake reached by a two-mile trail off rough Forest Road 626 east of Lizard Head Pass. It is worth an afternoon hike, as the view southeast from the Alta Lakes encompasses massive Mt. Wilson and Lizard Head silhouetted against the setting sun.

White River National Forest

Location: 2,253,000 acres in northwestern Colorado surrounding Vail, Aspen, Breckenridge, and Glenwood Springs. Interstate 70, U.S. Highway 24, and State Routes 82, 131, and 133 penetrate the forest. Maps and brochures available at the district ranger stations in Aspen, Eagle, Frisco, Meeker, Minturn, and Carbondale or by writing the Forest Supervisor, Old Federal Building, Box 948, Glenwood Springs, CO 81602.

Facilities and Services: Campgrounds (some with camper disposal service); picnic areas with grills; bus shuttle service to Maroon Bells; boat ramp at Ruedi Reservoir.

Special Attractions: Maroon Bells–Snowmass Wilderness; Flat Tops Wilderness; Eagles Nest Wilderness; Hunter-Fryingpan Wilderness; Collegiate Peaks Wilderness; Holy Cross Wilderness; Raggeds Wilderness.

Activities: Camping, picnicking, hiking, backpacking, hunting, fishing, boating, kayaking, horseback riding, nature study, spelunking, winter sports.

Wildlife: Elk, black bear, bobcat, mountain lion, bighorn sheep, mountain goat.

THE WHITE RIVER NATIONAL FOREST is a wild land. Seven different wilderness areas have been designated in this forest, and some of them contain the most awesome natural scenery in the country.

The Maroon Bells–Snowmass Wilderness tops the list, with its massive red-hued Maroon Peaks which rise 4,500 feet above cool, blue Maroon Lake—an unsurpassed show in contrasting beauty. You can either view the Maroon Peaks from the lake, or because of the popularity of the region, you can ride a shuttle bus the forest service operates to and from the area. You can get closer to the real thing by hiking from Maroon Lake to Crater Lake and then following West Maroon Creek to the peaks. At the extreme eastern edge of the wilderness, and just south of Cathedral Peak, is Conundrum Hot Springs. Here, at a cool elevation of 11,120 feet and near timberline, a small spring on Conundrum Creek pours forth warm water suitable for bathing. There is also a rare geological phenomenon known as a "knife edge" in this wilderness, but its appeal should only be to the serious mountain climber. This is a rock ridge not even two inches wide with sheer dangerous drop-offs on either side to ledges and cliffs below. It is located on the northeast route to Capitol Peak. Do not try this approach to Capitol Peak unless you are professional and experienced.

A different kind of wilderness is Flat Tops in the northern section of the White River National Forest. A small part of this wilderness is in the Routt National Forest. It contains high plateaus that are cut into, resulting in a giant structure that resembles a rock amphitheater out of Roman times. The most notable rock formation is called the Chinese Wall. There is a trail that parallels the wall on the east side, beginning at Trappers Lake and passing between Coffin and Little Trappers lakes before skirting in front of the wall. There are also dozens of clear, trout-filled lakes in the wilderness.

Another must on any itinerary is Eagles Nest Wilderness which is bordered for a while by Interstate 70. About half of the wilderness is in the Arapaho National Forest. There are several mountain peaks in excess of 12,000 feet; all except Bald Mountain and The Spider are in the Arapaho. A pleasant hike is from the Gore Creek Campground next to Interstate 70 (exit before Vail) to Deluge Lake, a distance of less than four miles. From the campground you can also follow Gore Creek due east across Red Buffalo Pass and into the Arapaho National Forest.

Not far away are two more outastanding forest wilderness areas: Hunter-Fryingpan Wilderness to the north and Collegiate Peaks Wilderness to the south. State Route 82 from Leadville to Aspen eventually snakes its way above tree line to Independence Pass, and at the observation site you are on a narrow strip of land between the two wild areas. At your feet is a good sample of alpine vegetation. During July and August dozens of showy wildflowers, many growing from matted stems, bedeck the mountaintop with bright red, blue, yellow, and white colors. As you descend the western slope from Independence Pass on Route 82, watch for the dilapidated log buildings from the abandoned mining town of Independence just before reaching the Lost Man Campground. This was an old mining town. As you complete your drive to Aspen, you will find yourself enclosed on either side by these two vast wildernesses. In addition, the Holy Cross and Raggeds wildernesses in the forest offer more backcountry adventures.

One other highly recommended area is Glenwood Canyon which was carved by the turbulent Colorado River and which is paralleled by U.S. Highway 6. The 16 miles between Glenwood Canyon and the Flat Tops Wilderness, bisected by Forest Road 600, is a country of sparkling crystal lakes, clear springs, rushing creeks, and an assortment of red-hued mountains. Many hiking trails and four-wheel-drive roads traverse the area.

Ashley National Forest

Location: 1,288,000 acres in northeastern Utah, north of Vernal. U.S. Highway 191 and State Route 44 pass through the forest. Maps and brochures available at the district ranger stations in Duchesne, Dutch John, Roosevelt, and Vernal, at the visitor's centers at Flaming Gorge and Red Canyon, or by writing the Forest Supervisor, 437 E. Main St., Vernal, UT 84078.

Facilities and Services: Campgrounds (most with camper disposal service); picnic areas with grills; boat ramps; visitor's centers; lodges, concessions.

Special Attractions: Flaming Gorge National Recreation Area; Red Canyon; Sheep Creek Geological Area; High Uintas Primitive Area.

Activities: Camping, picnicking, hiking, backpacking, hunting, fishing, boating, rafting, waterskiing, horseback riding.

Wildlife: Bighorn sheep, elk, moose, black bear.

WHEN YOU SEE this forest's special attraction of Flaming Gorge through which the Green River passes, you will experience the same sensation that the early explorer John Wesley Powell felt in 1869 when he wrote, "The Green River enters the range by a brilliant red gorge. We name it Flaming Gorge." Flaming Gorge is still as spectacular as it ever was, but there have been some changes since Powell discovered it. A 455-foot-tall dam was completed in 1963 across the Green River to form the huge Flaming Gorge Reservoir, and many recreation facilities have been built around the lake. Still, this has not detracted too much from the beauty of the area. We suggest that you start your exploration of the forest at the visitor's centers at Flaming Gorge Dam or at Red Canyon. There are good exhibits and literature about the canyons and gorges, the Uinta Mountains, and the area wildlife. There are several scenic overlooks into Red Canyon that provide unforgettable memories of fire-red rocks. Numerous campgrounds are in the area, as well as lodges at Flaming Gorge and Red Canyon. Boating and waterskiing are popular on the reservoir. For sheer excitement but safe enough for the entire family, save a half day for a float trip down the Green River from below the Flaming Gorge Dam to Little Hole. It is a white-water experience you will never forget.

A short distance west of Red Canyon is the Sheep Creek Canyon Geological Area, one of the most spectacular regions of upturned rocks in the country. You can hike or drive through this special canyon, and even examine some of the rocks for fossils, but it is *illegal* to take material from this area. The Palisades Picnic Area is an ideal place to lunch while in the canyon; Carmel Campground, overtopped by sweet-scented trees, is the place to spend the night.

The western section of the Ashley (along with a small eastern part of the Wasatch-Cache National Forest) contains the High Uintas Primitive Area. This region represents the only major mountains running east-west in all of the United States. Utah's highest mountain, Kings Peak at 13,528 feet, is located here, as well as many other mountains over 10,000 feet. More than half of the primitive area is bare rock with a minimum of vegetation. Below the timberline, however, particularly in the large, broad basins, are forests of Engelmann spruce, Douglas fir, subalpine fir, white fir, lodgepole pine, and limber pine. One special feature of the High Uintas is the opportunity to see a wide variety of birds. The water ouzel, which continuously flits around the streams, may be seen here. There are also water pipits, pine grosbeaks, Townsend's solitaires, and arctic three-toed woodpeckers.

The Wandin Campground at the southeast edge of the primitive area has trails that connect with the Highline Trail, which crosses Anderson Pass in the Kings Peak area. The trails are long and arduous, but crisp mountain air and dazzling wild scenery more than make up for the hike. No matter which trails you take, there is bound to be a wealth of clear, soothing lakes along them.

Dixie National Forest

Location: 1,900,000 acres in southwestern Utah, in the vicinity of Cedar City, Panguitch, and Escalante. Interstate 15, U.S. Highways 89 and 91, and State Routes 12, 14, 18, and 54 are the best access routes. Maps and brochures available at the district ranger stations in Cedar City, Escalante, Panguitch, St. George, and Teasdale or by writing the Forest Supervisor, 82 N. 100 E St., Cedar City, UT 84720.

Facilities and Services: Campgrounds (some with camper disposal service); picnic areas with grills; boat ramps, swimming beach.

Special Attractions: Remarkable red rock scenery.

Activities: Camping, picnicking, hiking, backpacking, hunting, fishing, boating, swimming, rock hounding, winter sports.

Wildlife: Mountain lion, bobcat, black bear, elk.

DIXIE is in the south—the south of Utah, where it is hot, dry, and exceptionally colorful. Boosters of tourism call this "Color Country," and it certainly is that. We love to stand on Strawberry Point and drink in the vivid reds and oranges of the canyon walls that extend for miles and miles. This experience is just the beginning of an adventure in that part of the Dixie southeast of Cedar City known as the Markagunt Plateau. From Strawberry Point backtrack to State Route 14. If you plan to be in this region for a while, make camp at the Duck Creek Campground near Route 14. From the campground hike or drive west for five miles to astonishing Navajo Lake. This lake was formed when a valley was dammed at one end by a lava flow; lava is still present around the lake area, particularly to the north. Beneath Navajo Lake is a series of lava tubes that were formed during a volcanic eruption. They act as pipes, draining water from the lake. Some of the lake water drains out at Duck Creek Spring and fills the pond you see at the Duck Creek Campground. Other water drains out of the side of a red limestone cliff, forming breathtaking Cascade Falls. The rocky trail to the falls follows a narrow ridge and passes several specimens of ancient bristlecone pines. There is a half-mile nature trail near Duck Creek Spring that wanders through a forest of ponderosa pine, white fir, Douglas fir, blue spruce, and quaking aspen. You will see extensive tree damage in this area; it is the result of porcupines who nibble on the tender branches. Nearby is an opening in the rocks known as the Ice Caves; here snow and ice that accumulate in heavy amounts during the winter linger through the summer.

At the western edge of the Markagunt Plateau, rising to 11,315 feet, is staggering Bryan Head Peak. The road or trail to the summit, where there are unsurpassed views into the red limestone country of Cedar Breaks National Monument, passes outcrops of shiny multicolored rocks containing sparkling quartz crystals. North of the peak is one of the most memorable picnic areas in the United States. Nestled beneath flaming red cliffs and located along a clear, deep blue stream is the Vermilion Castle Picnic Area. Some of the red formations resemble medieval castles. There are many other interesting formations to explore in this area.

To reach the Red Canyon area, one must go southwest of Panguitch. There is a comfortable campground along State Route 12, a good base for hiking and exploring among the red rocks. The Tropic Reservoir, south of Red Canyon, and near King Creek Campground, is one place to visit and wander among ponderosa pines.

The easternmost division of the Dixie is near Escalante. It is a good idea to start your explorations by hiking to Powell Point on the southernmost tip of Table Cliff Plateau. From this bristlecone-pine-bordered point, it is possible to see five states; however, there are no guard rails around the tip of this precipice and it is imperative that caution be observed. Hell's Backbone is near here and is part of an old mule trail that is now passable by vehicle following the construction of Hell's Backbone Bridge near Rogers Peak north of Escalante. This is a rough part of the forest, but the drive or hike is exciting and will send a few tingles down your spine. Blue Spruce Campground near here is set among handsome spruces and a clear mountain stream. The Pine Valley District of the Dixie, between Cedar City and St. George, includes Pine Valley Mountains, and vegetation ranges from desertlike rocky canyons to chilly high-altitude spruce forests.

Fishlake National Forest

Location: 1,424,000 acres in south-central Utah, between Panguitch and Nephi. Interstates 15 and 70, U.S. Highways 50 and 89, and State Route 10 are the access roads. Maps and brochures available at the district ranger stations in Beaver, Fillmore, Loa, and Richfield or by writing the Forest Supervisor, 115 E. 900 North, Richfield, UT 84701.

Facilities and Services: Campgrounds (most with camper disposal service); picnic areas with grills; boat ramps.

Special Attraction: Fish Lake.

Activities: Camping, picnicking, hiking, hunting, fishing, boating, nature study, horseback riding, winter sports.

Wildlife: Elk, bear, mountain lion, bobcat.

FISHLAKE takes its name from a large lake in the forest that suggests the outline of a giant fish. You can see the shape either from the air or by looking at a map, but it is not detectable, as you can visualize, at ground level. Fish Lake is a good place to begin exploration of the forest. At 8,800 feet, this 2,600-acre lake is lined with quaking aspen trees. The natural dam at the north end was formed by a glacier. Several campgrounds along the lake are convenient for frying the sumptuous rainbow and mackinaw trout that you will probably want to try a hand at catching. You do not have to confine your fishing to Fish Lake, however, since many of the small streams in the forest, such as Beaver River and Gooseberry Creek, are teeming with fish. In addition to the sport of fishing, you will find two scenic hiking trails to the Fish Lake Hightops from the west side of the lake. Doctor Creek Trail, from the Doctor Creek Campground, and Pelican Canyon Trail, from Bowery Haven, can be hiked easily in a day. Between Fish Lake and Interstate 70 which slices across the forest 25 miles to the north is a scenic region of canyons, valleys, plateaus, and high mountains. One area known as The Rocks has an amusing hodgepodge of huge rounded boulders. It can be reached by a route off Forest Road 50.

By contrast, on the western side of the forest, west of the Sevier River, are the high mountains of the Tushar Range. Birch Mountain, Circleville Mountain, Mt. Holly, Deer Trail Mountain, Gold Mountain, Mt. Baldy, and Delano Peak are all more than 10,000 feet, with Delano Peak reaching 12,173 feet. At the colder, higher elevations are Engelmann spruce, Douglas fir, alpine fir, and ponderosa pine. The Rydberg milk vetch, an endangered plant, grows on one of these high mountains. Plant lovers will also enjoy Partridge Mountain where few humans have interfered with the botanical life. There is a variety of special, seldom disturbed plant communities. The mountain itself peaks at 8,000 feet and is reached by a trail from South Walker Canyon that takes you through country of deep, rugged canyons, rocky ridges, and broad, grassy basins. In the moister canyons on the north side of the mountain are stands of white, subalpine, and Douglas firs. On at least one slope is an undisturbed community of bigtooth maple, mountain snowberry, and Gambel oak. Around the ridgetop to the east is a rocky area where Utah juniper, single-leaf pinyon pine, and sagebrush dominate. Downslope from them, in very dry, rocky soil, are mountain mahogany and cacti. The mountain can be explored in a day, but it is strenuous going in places.

Manti-LaSal National Forest

Location: 1,238,000 acres in central and southeastern Utah, in the vicinity of Price, Moab, and Monticello, with a small extension into western Colorado. U.S. Highways 50, 89, and 160 are access roads. Maps and brochures available at the district ranger stations in Ephraim, Ferron, Moab, Monticello, and Price or by writing the Forest Supervisor, 599 West Price River Drive, Price, UT 84501.

Facilities and Services: Campgrounds (some with camper disposal service); picnic areas with grills; boat ramps.

Special Attractions: Largest two-leaf pinyon pine in the world; hanging gardens, unique plant communities.

Activities: Camping, picnicking, hiking, hunting, fishing, boating, nature study, horseback riding.

Wildlife: Elk, black bear, mountain lion, bobcat, rare Navajo Abert squirrel.

WE HAVE SEEN every national forest, but we instantly fell in love with the Manti-LaSal as soon as we saw Fisher Towers and Warner Lake within 15 miles of each other. In approaching the forest we crossed the Colorado River on a one-lane suspension bridge and were following a narrow, deep gorge when Fisher Towers, a cluster of brilliant red pinnacles, came into view. As we got nearer we could detect the individual points of these rocky spires rising from the flat terrain. We then picked up Forest Road 62 at Castleton and began a winding journey through Pinhook Valley, on to Miners Basin, and to the top of a mesa where there were more views of the spectacular red rock country. We next descended to take the 5-mile side road to Warner Lake. It was here that we fell in love with the forest all over again. Warner Lake is situated in a stately forest of white-trunked aspens. Overhead is 11,612-foot Haystack Mountain, whose peak is reflected in the lake. It is the kind of breathtaking scene one finds in glossy nature magazines. To make it complete, the campground nearby is both remote and peaceful. From it are a choice of scenic hiking trails to Mill Creek, Oowah Lake, and Burro Pass.

Another attraction in this part of the forest is the largest two-leaf pinyon pine tree in the world near the North Beaver Road east of the Fisher Towers area. Its trunk is nearly 12 feet in circumference. Get precise directions to it from the district ranger.

A special distinct section of the Manti-LaSal worth a visit is west of Monticello and has several star attractions. For example, there are unique wet areas that harbor an interesting assemblage of plants such as a rare primrose, a death lily, and a variety of prairie-type grasses that includes little bluestem and side-oats grama. These areas are referred to as hanging gardens because they lie above the main valley. One accessible hanging garden in Arch Canyon can be reached by a trail from Forest Road 24 about 18 miles west of Blanding. A few miles northwest, in Lower Allen Canyon, is an area called Cliffdwellers Pasture, where pristine canyon vegetation of ponderosa pine, mountain snowberry, and Gambel oak combines with a natural arch, a natural bridge, and an assortment of archeological ruins to provide a more interesting, wild-looking garden area. Any artifacts are protected by law. Lower Allen Canyon is accessible from Forest Road 95 southwest of Mt. Linnaeus. West of Monticello are the only places in the world for the rare Navajo race of the Abert squirrel, an animal with a large, bushy white tail and large tufted ears.

The Manti division of the forest is 150 miles northwest of Monticello between the towns of Price and Manti. This is a region of higher, cooler mountains, scenic but rugged in character. North of Mt. Baldy, about a quarter mile north of Forest Road 22 from Mayfield, is Grove of the Giant Aspens, named for an outstanding group of quaking aspens, many with trunk diameters of 24 to 38 inches. Even though some of these denizens are showing signs of decay, they are still an impressive sight. Three miles east of the giant aspens, Forest Road 22 intersects with Skyline Drive, the rough but scenic route that follows the crest of the mountains for more than 70 miles. Since the road is often bad in places, it is best to check conditions with the district ranger. For those interested in plants, a journey to Scad Valley along the Left Fork of Huntington Canyon is imperative. There is a small cold water spring with alpine plants such as the prostrate gentian, alpine meadow rue, and several dwarf sedges, all species generally not found in this part of Utah.

Uinta National Forest

Location: 812,000 acres in central Utah, mostly south and east of Provo. Interstate 15, U.S. Highways 6 and 40, and State Routes 92 and 132 are the major roads. Maps and brochures available at the district ranger stations in Heber, Pleasant Grove, and Spanish Fork or by writing the Forest Supervisor, 88 W. 100 North, Provo, UT 84603.

Facilities and Services: Campgrounds (most with camper disposal service); picnic areas with grills; boat ramps; swimming beach.

Special Attractions: Mt. Timpanogos Scenic Area; Lone Peak Wilderness.

Activities: Camping, picnicking, hiking, backpacking, hunting, fishing, boating, swimming, winter sports.

Wildlife: Elk, black bear, moose, Rocky Mountain bighorn sheep.

MT. TIMPANOGOS towers mightily above the surrounding terrain about 35 miles southeast of Salt Lake City. It is in the heart of the Uinta's Mt. Timpanogos Scenic Area, one of the forest's special attractions. A paved highway encircles part of the scenic area so that those who wish to enjoy it from the outside can do so, but to get the feeling of "Timp," as the peak is nicknamed, hiking into the interior is the way to go. Our favorite way is via the Timp Trail that leaves from the Aspen Picnic Area along State Route 50. It is 7 miles to the summit of Mt. Timpanogos; en route you pass a huge glacier, skirt around sparkling Emerald Lake, and arrive at the summit house at 11,750 feet. As soon as you catch your breath, look in all directions at the vast panoramas laid out before you. By starting early in the morning, it is possible to hike to Timp and back in one day.

Lone Peak, the other outstanding attraction in the forest, stands isolated above Upper Box Elder Canyon whose steep slopes are covered with sad-looking windswept firs. The peak is the focal point of the 30,088-acre Lone Peak Wilderness, a rugged land mass shared by the Uinta and Wasatch-Cache national forests. Since the elevation range in the wilderness is from 5,000 to more than 11,000 feet, the biological life zones are fascinating and diverse. The lower elevations on the western slopes are home to typical dry vegetation such as sagebrush, bitterbrush, and an assortment of grasses. Upslope, in the more moist areas, are Gambel oak, Rocky Mountain maple, and ninebark. Beyond that, in the cooler climes, aspens and white fir take over, then Douglas fir, and finally, at the chilliest, highest elevation where trees can still grow, alpine fir and Engelmann spruce predominate. Above timberline, bare patches of rock alternate with low, matted vegetation.

We also recommend the Midway-Alpine Loop Road which circles the south side of Deer Creek Dam and winds its way across canyon walls before dropping down into Deer Creek Canyon and the fabulous Cascade Springs. The springs gush out and drop from a series of ledges that form a small horseshoe-shaped amphitheater. Several rustic footbridges add a touch of nostalgia to the area.

Another worthwhile loop trip follows the Right Fork of Hobble Creek out of Springville. Just beyond Jux Canyon the route turns south and parallels Diamond Fork Creek until it reaches Spanish Fork Canyon. This canyon can be followed back to Springville. About halfway around, near the Mineral Spring Campground, you will discover an open, marshy area with several kinds of showy wildflowers.

Mt. Nebo is another prominent landmark in the forest. At 11,877 feet, it is more than 100 feet higher than Mt. Timpanogos. It can be reached by several good trails, or it can be viewed from the Mt. Nebo Scenic Loop Road which leaves Nephi Canyon east of Nephi. If you take this road, stop and hike the short trail to Devil's Kitchen, an area of eroding red sandstone formations. Should you then continue on the Mt. Nebo Road, you can exit either through Santaquin or Payson Canyon, both worth the scenic journey. The Payson Canyon route passes the popular Payson Lakes.

Another junket we suggest is to Provo Canyon, a beautiful gorge carved by the Provo River. Famous Bridal Veil Falls is nearby, off U.S. Highway 189 which follows the canyon.

Wasatch-Cache National Forest

Location: 1,303,000 acres in northeastern and north-central Utah, from Logan to Salt Lake City and also east of Heber City; also two small units west of Salt Lake City and Provo as well as an extension into Wyoming. Interstates 15, 80, and 84, U.S. Highway 91, and State Routes 39 and 150 are the major access roads. Maps and brochures available at the district ranger stations in Kamas, Logan, Ogden, and Salt Lake City, Utah, and Evanston and Mountain View, Wyoming, or by writing the Forest Supervisor, 125 S. State St., Salt Lake City, UT 84138.

Facilities and Services: Campgrounds (some with camper disposal service); picnic areas with grills; boat ramps.

Special Attractions: High Uintas Primitive Area; Lone Peak Wilderness; Jardine Juniper, largest Rocky Mountain juniper in the world.

Activities: Camping, picnicking, hiking, backpacking, hunting, fishing, boating, waterskiing, horseback riding, winter sports.

Wildlife: Black bear, bobcat, mountain lion, elk.

THE TWO SPECIAL ATTRACTIONS of this forest are also found in the Ashley and Uintas forests; a visit to all three forests is highly recommended if you have the time. Let us look at the first highlight of the Wasatch, the High Uintas Primitive Area, which it shares with Ashley Forest. A drive along State Route 180 from Kamas to the Wyoming border will give you a preview of the area. From the highway you can see Smooth Rock Falls and rugged Slate Gorge, peer down at the thundering Provo River Falls, and climb to Bald Mountain Pass where you can gaze at the High Uintas Primitive Area at closer range. Spectacular Mt. Agassiz and Hayden Peak loom in the distance, and Mirror Lake, so-called because of its silvery luster that reflects the mountain peak, glistens below. To the right of Mt. Agassiz is Naturalist Basin. If you have time, take the eight-mile trail from Mirror Lake to the basin, which contains several lakes surrounded by rich wetlands. The birds, other animal life, and plants common to a wetland habitat make this truly a naturalist basin, an area with a great deal of diversity of living things.

South of Little Cottonwood Canyon is another highlight of the forest—the beginning of Lone Peak Wilderness, a rugged, high mountain region that extends into the Uinta Forest. This wilderness contains Bells Canyon, Upper Bells Canyon Reservoir, and Little Matterhorn Peak, a prominent mountain with a pointed peak.

Another part of the forest is the Cache division which lies north and east of Ogden and Logan and continues to the Idaho state line. (The part of the Cache that extended into Idaho is now part of the Caribou National Forest and is described in that entry.) The Cache includes the north end of the Wasatch Range, where Mt. Ben Lomond and Mt. Ogden reach 9,712 and 9,572 feet, respectively, and the Bear River Range, through which spectacular Logan Canyon has been carved. Skyline Trail, a good hiking trail from the campground at Willard Basin to Mt. Ben Lomond and other high mountains to the south, passes scenic canyons and bubbling springs and offers exquisite panoramic views.

Twenty-five miles to the north, Logan Canyon begins its mile-deep cut through fossil-bearing limestone, shale, and sandstone. There is plenty of adventure along Logan Canyon. From the DeWitt Campground is a mile-long trail to Wind Cave, which has been formed by intense wind etching into soft limestone. There is a one-and-a-half-mile steep route to the largest Rocky Mountain juniper in the world from Wood Campground. Known as the Jardine Juniper, this forest attraction has a trunk diameter of 8 feet and is estimated to be 3,000 years old. Other nearby attractions are Logan Cave, Hidden Spring, and the fast-gushing Ricks Spring. The road and trail begin to climb out of Logan Canyon on their way to the Bear Lake Overlook. Stop or stay at the Sunrise Campground and take the well-worn nature trail to a nearby record-sized limber pine which has a trunk circumference of nearly 22 feet. On the way to it are communities of firs, spruces, pines, aspens, and mountain mahoganies.

Some of the best places to hike in the forest are along the three big canyons—Mill Creek, Cottonwood, and Little Cottonwood—that have been carved into the mountains east of Salt Lake City. A network of trails connects such places as Grandeur Peak, Mt. Olympus, Gobbler's Knob, and Desolation Lake. Most of these destinations are within a day's hike of the campgrounds in the three canyons.

Humboldt National Forest

Location: 2,527,000 acres in northern and east-central Nevada. Interstate 80, U.S. Highways 6, 50, 93, and 95, and State Route 225 pass through the forest. Maps and brochures available at the district ranger stations in Ely, Mountain City, Wells, and Winnemucca, Nevada, and Buhl, Idaho, at the Lehman Caves National Monument Visitor's Center, or by writing the Forest Supervisor, 976 Mountain City Highway, Elko, NV 89801.

Facilities and Services: Campgrounds (some with camper disposal service); picnic areas with grills; visitor's center.

Special Attractions: Jarbidge Wilderness; Wheeler Peak Scenic Area; stands of ancient bristlecone pines; Ruby Mountains Scenic Area.

Activities: Camping, picnicking, hiking, backpacking, hunting, fishing, nature study, horseback riding, swimming (at Angel Lake).

Wildlife: Threatened bald eagle; mountain lion, Rocky Mountain bighorn sheep, elk, golden eagle.

THE HUMBOLDT is an area of extremely rough mountains, deep canyons, and rounded basins scoured by glaciers aeons ago. There are nine distinct units of the forest, including some that extend to the Oregon, Idaho, and Utah borders.

The wildest and most unspoiled land is found in one of its star attractions, the 64,000-acre Jarbidge Wilderness. The area has a dazzling assortment of sculptured rock formations that the elements have been working on for aeons. It also boasts of a wide variety of habitats that run the gamut from dry desert areas at the lower elevations where sagebrush thrives to lofty rock peaks where spruce, fir, and pine forests thrive. Another attraction is the golden eagles usually seen soaring overhead. There is a network of challenging trails that penetrate the wilderness.

Our favorite area, however, is the Wheeler Peak Scenic Area west of Baker near the Utah state line. The Lehman Caves National Monument, at the eastern edge of this scenic area, has a visitor's center operated jointly by the National Park Service and the Humboldt Forest. It is well worth a visit to learn about the staggering number of sights in the Wheeler Peak area. To capture the highlights, it is suggested that one drive the circuitous but superbly scenic road starting at Lehman Caves to the Wheeler Peak Campground. From here there is a short trail to Teresa Lake. If it is summer you will see the fragile, pink Parry's primrose and the yellow Nevada primrose in bloom, as well as the vivid scarlet monkey-flower. South of Teresa Lake, branch off to the Bristlecone-Icefield Trail. For two miles the trail climbs up rocky slopes through a stand of ancient bristlecone pines that dates back hundreds of years. Pause for a moment to see their twisted trunks and to note the limited amount of living material in these trees. Continue on into the ancient glacial basin that contains a permanent ice field. The stark, awesome beauty of this bleak, icy region will overwhelm you. Unless you are an experienced mountain climber, *do not* attempt to scale the summit of the 13,063-foot Wheeler Peak. But there are other rewards. About one and a half miles southeast of the scenic area, reached by hiking a short distance from the end of the extremely rough Lexington Creek Road (Forest Road 450), is Lexington Arch, an unusually tall limestone arch about six stories high.

Another area full of exciting sights is the Ruby Mountains Scenic Area southeast of Elko. One can enter through the dramatic and colorful Lamoille Canyon. Nearby where Thomas Canyon branches south from Lamoille Canyon is a perfect base and campground for exploring the Ruby Mountains. There are dozens of secluded lakes and sparkling streams. Overland Lakes, set among intensely rocky terrain, are in the remote southwestern corner. Only time and strength are your limitations.

For a special historic tour of the forest, take U.S. Highway 6 southwest from Ely for about 15 miles. Then hike or drive the old Hamilton Pioche Stage Route (Forest Road 402) to the site of a stage station, several natural springs, and the ruins of Hamilton, an old mining town. Beyond Hamilton the route follows the Mokomoke Mountains until it emerges at U.S. Highway 50. About 40 miles east of Ely is the Table, a one-square-mile, 11,000-foot flattopped plateau east of Mt. Moriah. This unusually large plateau supports a stand of ancient bristlecone pine and a myriad of colorful, summer-blooming wildflowers. It is also a good place to spot a Rocky Mountain bighorn sheep, distinguished by its unusually large curved horns. As you climb to the Table from the end of Forest Road 582 along Hampton Creek, you will encounter juniper, pinyon pine, ponderosa pine, limber pine, white fir, and aspen.

North of Winnemucca are unusual rocky pillars that rise 200 feet into the air near State Route 88 along Indian Creek from Paradise Valley to Hinkey Summit. Although only 7,865 feet above sea level, this vantage point offers a sweeping view of the Santa Rosa Range to the west.

Toiyabe National Forest

Location: 3,193,000 acres in central, western, and southern Nevada, with a small extension into eastern California. Interstate 80, U.S. Highways 50 and 95, and State Routes 338, 559, 378, and 157 serve the forest. Maps and brochures available at the district ranger stations in Austin, Bridgeport, Carson City, Las Vegas, and Tonopah, or by writing the Forest Supervisor, 111 N. Virginia St., Reno, NV 89501.

Facilities and Services: Campgrounds (some with camper disposal service); picnic areas with grills; boat ramps; swimming beach.

Special Attractions: Carson-Iceberg Wilderness; Hoover Wilderness; deserts with Joshua trees.

Activities: Camping, picnicking, hiking, backpacking, hunting, fishing, boating, swimming, horseback riding, nature study.

Wildlife: Endangered or threatened bald eagle, American peregrine falcon, and Paiute trout; desert bighorn sheep, mountain lion, elk, bobcat, antelope.

TOIYABE FOREST is the largest of all national forests outside of Alaska. It is divided into three major parts, each varying geologically and biologically from one another.

The westernmost Sierra division is on the eastern front of the massive Sierra Nevadas. The topography of this section ranges from the steep, serrated peaks of Sawtooth Ridge to the rounded domes of Sweetwater Mountains. At the upper end of this part of the forest is famous Lake Tahoe, one of the most beautiful of western lakes, surrounded by three different national forests. The Toiyabe is responsible for maintaining the popular beach area on Lake Tahoe's southeast shore. There are a pleasant campground, picnic site, and swimming beach at this location. A few miles northeast of Lake Tahoe, and reached by scenic State Route 27, is the Mt. Rose Campground. It offers a good view of Lake Tahoe, but by hiking a few miles north to the 10,778-foot summit of Mt. Rose, you will also have an overpowering panoramic vista of Lake Tahoe to the south, Reno to the northeast, and Truckee Canyon to the northwest.

Most of the Sierra division lies southeast of Lake Tahoe on the California side of the state line. The 92,000-acre Carson-Iceberg Wilderness and the 48,000-acre Hoover Wilderness are part of this area. The Carson-Iceberg is a mountainous region with deep canyons, thundering waterfalls, high meadows, and clear streams. Even though the Rodriquez Flat Trail Head is the most frequently used route and often crowded, we like it best. The Driveway Trail (#1038) begins here by climbing straight up through a dense forest of lodgepole pine, silver pine, and white fir. En route you will discover an interesting stack of rocks called an ari mutillak. It resembles a chimney and was built in the 1920s by Basque sheepherders, who incidentally form a good part of the population in western states, having immigrated here to work as sheepherders. Although the Driveway Trail continues for three miles to the Silver King Creek Trail, we suggest that you branch south on the Corral Valley Trail (#1039) as it leads to a gigantic, rare, double-stemmed juniper tree whose two trunks together are 12 feet in diameter and 36 feet in circumference. It is quite a sight to behold.

Other not-to-be-missed points of interest in the Carson-Iceberg are Tamarack Lake, inhabited by golden trout and surrounded by sheer cliffs, the pounding Llewellyn Falls, and the volcanic dome known as White Soda Cone whose crater bubbles with ice-cold water.

Not far away is Hoover Wilderness, another special forest site, which lies adjacent to the northeastern corner of Yosemite National Park. It has broad canyons with shallow streams bordered by grassy meadows in between rugged mountains. You will have an opportunity to see a considerable number of lodgepole, limber, and Jeffrey pines and western hemlocks in this area. Matterhorn Peak, which rises to 12,264 feet in the Sawtooth Ridge, sports several patches of glaciers. Farther south, between Walker and Mono lakes, is an isolated part of the Toiyabe that includes the biologically unique Alkali Valley, where plants can tolerate an alkaline soil that is more extreme than any other place in the world. Be sure to take plenty of water or other liquid refreshment before hiking in this arid but lovely place.

The second part of the Toiyabe is the Las Vegas district,

a small but mighty area less than 50 miles west of Las Vegas and dominated by 11,918-foot Charleston Peak in the Spring Mountain Range. This round-topped mountain is rimmed by ancient bristlecone pines. The outstanding feature of the Spring Range is that it is surrounded by desert, a highlight of the forest. To appreciate this, we suggest driving the road from Kyle Campground which connects State Routes 157 and 156 to the Desert View Scenic Overlook. Joshua trees, cholla cacti, agaves, and other desert vegetation thrive in this valley, silhouetted against a peak with the marvelous name of Mummy Mountain.

The third but huge central division of the Toiyabe is located smack in the middle of Nevada, north of Tonopah. It consists of three parallel mountain ranges (Toiyabe, Toquima, and Monitor), all in the forest, separated by two broad valleys—Big Smoky and Monitor—that are not in the forest. These desert mountains contain hundreds of natural springs, abandoned mines, and the remains of gold and silver towns. The mountains overlook the old Pony Express Trail, used by those rugged souls who carried the U.S. mail by horse, and the trail used by the 1843 Fremont expedition, which explored the unknown West. We suggest using the 72-mile Toiyabe Crest Trail. It is a rugged hiking trail that follows the high ridge of the Toiyabe Mountains from Kingston Canyon to South Twin River Canyon. Rare desert bighorn sheep can be seen occasionally wandering along this trail.

SOUTHWEST

The southwestern region of the United States contains 11 national forests. They are the Apache-Sitgreaves, Coconino, Coronado, Kaibab, Prescott, and Tonto in Arizona and the Carson, Cibola, Gila, Lincoln, and Santa Fe in New Mexico.

Apache-Sitgreaves National Forest

Location: 1,904,000 acres in northeastern Arizona, west and south of Springerville. U.S. Highways 60, 180, and 666 and State Routes 77 and 260 are the major highways. Maps and brochures available at the district ranger stations in Alpine, Clifton, Lakeside, Overgaard, Springerville, and Winslow or by writing the Forest Supervisor, Federal Building, Box 640, Springerville, AZ 85938.

Facilities and Services: Campgrounds (some with camper disposal service); picnic areas with grills; boat launch areas.

Special Attractions: Mogollon Rim; Blue Range Wilderness; Mt. Baldy Wilderness.

Activities: Camping, picnicking, hiking, backpacking, hunting, fishing, boating, horseback riding, nature study, winter sports.

Wildlife: Endangered or threatened bald eagle and Apache trout.

THE APACHE-SITGREAVES FOREST is dominated by the Mogollon Rim, one of the forest's most spectacular attractions. This rim is an almost continuous vertical cliff—in places it is as high as 1,000 feet—and separates Arizona's high country to the north from the rugged low country southward. The Apache part of the forest extends from Springerville south nearly to Clifton and is crossed by the Mogollon Rim.

One of the best ways to get an overview of the Apache section is to use the rim. To do this, take Route 666 from Alpine, known as the Coronado Trail since it loosely follows the Spanish explorer Coronado's route taken in 1540 in the Seven Cities of Cibola. This highway climbs to and descends from the Mogollon Rim in a series of sharp twists and turns. Although only about 70 miles long, the road usually requires four hours to drive *without stops*. Plan on stops at Hanagan Meadows where there is a lush meadow in an otherwise continuous forest of pines, firs, and spruces; at Blue Vista, an observation point with an unending view to the west; and at the Blue Lookout Tower which has a commanding view to the east over the vast Blue Range Wilderness.

The Blue Range Wilderness is another must attraction in the forest. It has 173,762 acres of rugged, rock-rimmed canyons, large areas of exposed rock surfaces, and unique box canyons that close in on three sides. The Mogollon Rim, which penetrates the wilderness, is forested on top with spruces and firs. Below the rim are ponderosa pines, pinyon pines, and western junipers. A prominent feature of the wilderness is the well-named Sawed Off Mountain, a mountain that looks as though someone took a handsaw to it. A good trail in the wilderness is the Blue Lookout Trail

(#71) which drops 2,200 feet from the tower to Blue Cabin, which once served as a residence for the Blue Lookout crew. By a number of switchbacks, the trail goes from a forest of aspen and pine to a mixture of conifers and hardwoods along KP Creek.

A much smaller wilderness, but also a forest special attraction, is Mt. Baldy Wilderness about 20 miles southwest of Springerville. This is Arizona's only wild area that contains a subalpine vegetation zone composed of Douglas fir, white fir, ponderosa pine, and spruce, with occasional dense stands of quaking aspens. Several meadows, filled in summer with wildflowers, are interspersed among the trees. There are breathtaking views in all directions from the summit of Mt. Baldy. The most impressive trail, and one that can be hiked to the top of Mt. Baldy and back in one day, leaves near Phelps Cabin on Route 73 and climbs to the summit of the mountain by way of the East Fork of the Little Colorado River. Most of the trail is through woodlands dominated by Engelmann spruce, blue spruce, Douglas fir, cork-bark fir, white fir, ponderosa pine, and southwestern white pine. The Phelps Cabin area, made up of spruce forests, Douglas fir forests, wet meadows, and grasslands, is considered by botanists to have one of the most diverse floras in Arizona.

Most of the activity in the Sitgreaves part of the forest is also along the Mogollon Rim. One of the largest stands of ponderosa pine in the world as well as sweeping vistas and trout-filled lakes are found here. A road, often primitive, follows the rim for nearly 75 miles from the town of Show Low to Black Lake, Willow Springs Lake, Woods Canyon Lake, and Bear Canyon Lake. Fishing and boating are great sport at all of these lakes.

For another treat, drive 18 miles from Heber to the Chevelon River and hike the Chevelon Canyon which extends for miles in either direction. There is a secluded campground where Forest Road 504 crosses the river and its canyon.

Popular winter activities in the Apache-Sitgreaves include snowmobiling, downhill skiing, cross-country skiing, and ice fishing.

Coconino National Forest

Location: 1,800,700 acres in central Arizona, surrounding Flagstaff. Interstates 17 and 40, U.S. Highways 89, 89A, and 180, and State Route 87 are the major roads. Maps and brochures available at the district ranger stations in Flagstaff, Happy Jack, Rimrock, and Sedona or by writing the Forest Supervisor, 2323 E. Greenlaw Lane, Flagstaff, AZ 86001.

Facilities and Services: Campgrounds (some with camper disposal service); picnic areas with grills; boat launch areas.

Special Attractions: Oak Creek Canyon; Sycamore Canyon Wilderness; San Francisco Peaks; Sinagua Indian Tribe ruins.

Activities: Camping, picnicking, hiking, backpacking, hunting, fishing, boating, horseback riding, nature study, winter sports.

Wildlife: Endangered or threatened bald eagle and American peregrine falcon; black bear, mountain lion.

INSPIRATIONS ARE BORN in the Coconino National Forest, and well they should be, since Oak Creek Canyon is one of the world's most vividly colored rock formations. The rocks lining it are tinted in exquisite shades of pink, red, and purple. Oak Creek Canyon was created when streams cut through the limestone and sandstone of the nearby Mogollon Rim, a continuous cliff that is over 1,000 feet in many places. Near Sedona, where Oak Creek flows through the Verde Valley, are the most brilliant fire-red buttes and cliffs in the country. The campground at Red Rock Crossing is in the heart of these sandstone buttes; it goes without saying that this is a mesmerizing spot to pitch camp. From this campground you can also see the spires of Cathedral Rock, which reflect in the water of Oak Creek. Hiking trails in the canyon pass several natural stone arches, including the high and narrow Devil's Bridge and an unusual formation known as Vultee Arch.

The trail that follows the West Fork of the canyon enters a rocky area of craggy cliffs. Along the stream are fine samples of box elder, walnut, bigtooth maple, Gambel oak, and the showy, pink-flowered New Mexican locust. As the elevation increases on the shady, north-facing slopes, a forest of Douglas fir and ponderosa pine takes over. Look for Abert's squirrels, golden-colored ground squirrels with large ears, and bright red and yellow western tanagers in this habitat. Across from the pines and firs, on the dry, south-facing slopes, is typical desert flora such as prickly pear cacti, century plants, manzanitas, mountain mahoganies, gnarly shrub live oaks, pinyon pines, and western junipers. If you can take your eyes off the rugged scenery long enough, you can fish for brown and rainbow trout in the West Fork.

For a never-to-be-forgotten, white-knuckle drive, take the twisting Schnebly Hill Road from Sedona and climb onto the Mogollon Rim. You will pass dazzling red rock formations, one named Merry-go-round Rock. You can also drive State Route 179 south to Bell Rock Vista, where you will be convinced that the whole world is a glowing red butte.

West of Oak Creek Canyon is another prize forest attraction—Sycamore Canyon Wilderness, which extends into the Prescott and Kaibab forests. Its focal point is the multicolored Sycamore Canyon which resembles a small-scale Grand Canyon. The canyon runs for 21 miles and spreads 5 miles from rim to rim at Sycamore Basin. Large sycamores, cottonwoods, and willows line its bottom, while ponderosa pine and Douglas fir grow along the rim at 7,000 feet. Black bears and mountain lions are here, as are javelinas and ring-tailed cats. We, unfortunately, did not happen to see any of these formidable creatures on a recent visit. Golden eagles sometimes soar and dive overhead. Rattlesnakes and scorpions are here, too, so keep this in mind.

In sharp contrast to the red and purple canyons and gorges are the San Francisco Peaks, another forest feature that gets top billing. The four giant peaks surround a lush green valley known as the Inner Basin. Tallest of these old volcanic mountains, at 12,633 feet, is Humphreys Peak, but the others—Agassiz, Fremont, and Doyle—are high enough to have tundra on their summits, where the vegetation is dwarfed, low-growing, and often matted. Because the ecology of these mountain summits is extremely fragile, visitors are requested not to leave established trails. Some of the last trees that form the timberline are gnarled and ancient bristlecone pines, the oldest kind of living organisms in the world, although younger, dwarf Engelmann spruces are also here. Showy wildflowers found near timberline include yellow columbine, purple gentian, silvery lupine, and the deep pink Parry's primrose. The Inner Basin, nearly surrounded by these four great peaks, is noted for its aspen groves and lush meadows.

Another big forest attraction is the dozens of traces of the prehistoric Sinagua Indian people. Among the remains of their lives are multiroomed pueblo dwellings on mesas, small caves with rooms in them, remains of a ball court and fields with crude irrigation systems, a masonry fort, numerous pictographs, or rock paintings, and much broken pottery. As you would expect, it is illegal to collect or molest any of these ruins. Check with any of the distinct rangers about the accessibility of these areas.

For recreation there is Mormon Lake, south of Flagstaff, but it has fluctuating water levels and is beginning to fill in with marshes. Hunting is good, but fishing and boating have become poor in recent years. At Slide Rock in Oak Creek Canyon, waders and swimmers can at least cool off in the clear, rushing water.

Coronado National Forest

Location: 1,790,900 acres in southern Arizona, from Tucson and Safford to the borders of New Mexico and Mexico. Interstates 10 and 19, U.S. Highways 80, 89, and 666, and State Route 82 are the major roads. Maps and brochures available at the district ranger stations in Douglas, Nogales, Safford, Sierra Vista, and Tucson, at the Sabino Canyon Visitor's Center, or by writing the Forest Supervisor, 301 W. Congress, Tucson, AZ 85701.

Facilities and Services: Campgrounds (some with camper disposal service); picnic areas with grills; boat launching areas; visitor's center; concessions.

Special Attractions: Diverse vegetation zones on Mt. Lemmon; unusual Mexican plants and animals in Sycamore

Canyon; Chiricahua Wilderness; Galiuro Wilderness; Pusch Ridge Wilderness.

Activities: Camping, picnicking, hiking, backpacking, hunting, fishing, boating, swimming, horseback riding, nature study, winter sports.

Wildlife: Endangered jaguarundi; very rare coppery-tailed trogon and Gila chub; desert bighorn sheep.

THE HITCHCOCK HIGHWAY east of Tucson climbs from the desert at Sabino Canyon to frigid forests on 10,720-foot Mt. Lemmon, a winter sports area. Stop at the visitor's center in Sabino Canyon to learn about the five life zones you will pass through on your way to the top. These zones are desert, grassland, pinyon-juniper, ponderosa pine woodland, and the Canadian zone. Then see for yourself where the desert foothills give way to grasslands which in turn give way to the woodland which gives way to ponderosa pines, until higher up everything gives way to the Canadian zone of Douglas fir and Engelmann spruce.

In addition to Mt. Lemmon there are several special attractions to explore in the Coronado. Our favorite is Sycamore Canyon, 15 miles west of Nogales and 7 miles southeast of the nearly abandoned town of Ruby. Forest Road 39, an often narrow and rough gravel road, crosses Sycamore Creek just north of Hank and Yank Spring, the site of a primitive campground. A short distance below the spring, Sycamore Creek has carved a harsh and formidable canyon that extends for more than 5 miles south before plunging into Mexico. The canyon is the only place in the United States where several kinds of plants and animals live. For starters, you can see (1) the rare Goodding ash tree; (2) the yew-leaved willow; (3) the Mexican tapioca, a shrubby plant closely related to the tropical cassava or tapioca; (4) the resurrection plant, related to Spanish moss; (5) the whisk-broom plant, a leafless plant related to ferns; (6) the golden-chested beehive cactus; and many, many other rarities. Unusual animals, some of which used to live in but may now be gone from Sycamore Canyon, include the pencil-thin vine snake, jaguar, ocelot, and coppery-tailed trogon, a magnificent bird of red, green, and white plumage. This area is also the only habitat in the United States for the jaguarundi, which is a handsome, slinky member of the cat family.

Another tantalizing area is Goudy Canyon in the Graham Mountains southwest of Safford. To get there, you must take the Swift Trail Highway, which leaves State Route 666, for 31 miles up the Mt. Graham Road. In some places the canyon's cliffs drop sharply from 9,000 to 7,000 feet in about 1½ miles. On the upper slopes is a mixed conifer stand with a high percentage of Mexican white pine. There are also splendid Douglas firs, white firs, ponderosa pines, aspens, and Gambel oaks.

Still another fabulous canyon in this area is Pole Bridge Canyon. It has an out-of-the-ordinary display of plants not to be missed that include Apache pine, Chihuahua pine, Arizona white oak, net-leaf oak, silver-leaf oak, several kinds of agaves, and bear grass. There are also a number of interesting birds, among them Stephen's vireo, Mearns woodpecker, Rocky Mountain towhee, and ruby-crowned kinglet. To reach the canyon, leave Douglas via Highway 666 to State Route 181, then go west for 12 miles before turning west again on Turkey Creek Canyon Road. Follow this gravel road for 10 miles until it crosses Pole Bridge Canyon Creek on a one-lane concrete bridge where the trail head is located. The trail runs about 2½ miles.

Madeira Canyon is another lovely forest area. The road to it from Continental leads through true desert country of barrel cacti, cholla, and ocotillo. At its right fork we found pleasant picnic sites and campgrounds along the stream. We only spent one night here, but wished we could have spent more. Some 200 kinds of birds have been recorded from Madeira Canyon, making it a paradise for bird-watchers. A hiking trail to towering Mt. Wrightson begins here.

To the east is Cochise Stronghold, the reported burial ground of Cochise, the famous Indian warrior. Today you can hear a pin drop in this area; it is a good place to see samples of the silver-leaf oak.

Near the New Mexican border is another treat: the beautiful, light brown cliffs of Portal. The road from the town of Portal up to the Chiricahua National Monument is lined with rugged rocks. Coppery-tailed trogons, unusual coatimundis, and Chiricahua fox squirrels are often seen about. This road is just north of the 18,000-acre Chiricahua Wilderness which occupies the summit of the Chiricahua Mountains. Some of the steepest and nearly inaccessible canyons in southeastern Arizona are in the wilderness, which helps distinguish it as a special forest attraction. On Chiricahua Peak is the southernmost location in the country for Engelmann spruce.

Another special forest wilderness area is the Galiuro Wilderness northwest of Willcox and northeast of Tucson. The area is known for its mountains, steeply and abruptly rising from the surrounding plains. Many of the tall cliffs are brightly colored. The wilderness has dense vegetation, and for that reason straying from the marked trails is prohibited.

If you want to glimpse rare desert bighorn sheep with their curved horns, go to still another forest attraction, Pusch Ridge Wilderness, which overlooks the city of Tucson. The ridge is home to a small population of these rare sheep. Pusch Ridge, located on the western flank of the Santa Catalina Mountains, has a number of canyons worth exploring.

In all this exciting desert land are some genuine oases. Among them are Pena Blanca Lake, Parker Canyon Lake, and Riggs Flat Lake. Boating and fishing are good sport at all of them.

Kaibab National Forest

Location: 1,558,650 acres in north-central Arizona, on either side of the Grand Canyon. Interstate 40, U.S. Highways 89, 89A, and 180, and State Routes 64 and 67 are the major highways. Maps and brochures available at the district ranger stations in Fredonia, Grand Canyon, and Williams or by writing the Forest Supervisor, 800 S. 6th St., Williams, AZ 86046.

Facilities and Services: Campgrounds (some with camper disposal service); picnic areas with grills; boat launching areas; concessions.

Special Attractions: Entrance routes to the Grand Canyon; Sycamore Canyon Wilderness.

Wildlife: Unusual Kaibab squirrel; endangered or threatened bald eagle and American peregrine falcon.

THE KAIBAB lies north and south of the Grand Canyon, and like the north and south rims of the canyon, the two units of the forest are dramatically different from each other.

The north part of the Kaibab is a cool forest of pine, spruce, and aspen, where the rare Kaibab squirrel, recognized by its dark body and white tail, can be seen scurrying about in the woods. The south part of the Kaibab is much drier, with pinyon and juniper the dominant plants; the Abert squirrel replaces the Kaibab squirrel. Bison and wild turkey may be seen in the north Kaibab, where both have been introduced in the area.

In the north Kaibab, Jacob Lake and DeMotte Park are the most popular campgrounds. They are along the entrance road to the north rim of the Grand Canyon. Another camping area is the small and remote Indian Hollow Campground, within a half mile of the Grand Canyon boundary. Reached by a rough road, the campground is within easy reach of the Thunder River Trail which penetrates Grand Canyon National Park. We found it a convenient spot to pitch camp, as we could make a variety of excursions from there into and around the north rim of the Grand Canyon. A few miles south of Jacob Lake, on the east side of Route 67, is Crane Lake. It is a subalpine mountain lake in a parklike setting. The edge of the lake is extremely marshy and supports a rich wetland flora.

Several lakes in the southern unit of the Kaibab are suitable for boating. We like White Horse Lake the best because of its beautiful setting and its superior fishing. There is also a short nature trail near the lake. Nearby, one of the more enjoyable hiking trails begins west of Williams and climbs to Bill Williams Mountain for an overview of the surrounding territory. A tortuous gravel road also leads to the summit.

The Sycamore Canyon Wilderness, which barely enters the Kaibab, is described in the Coconino and Prescott National Forest entries; it is one of the forest's drawing cards, and we recommend a side trip to it.

Prescott National Forest

Location: 1,236,900 acres in central Arizona on all sides of Prescott. Interstate 17, U.S. Highways 89 and 89A, and State Route 169 are the primary highways. Maps and brochures available at the district ranger stations in Camp Verde, Chino Valley, and Prescott or by writing the Forest Supervisor, 344 S. Cortez, Prescott, AZ 86301.

Facilities and Services: Campgrounds (some with camper disposal service); picnic areas with grills; boat launching areas.

Special Attractions: Sycamore Canyon Wilderness; Pine Mountain Wilderness; original stagecoach station.

Activities: Camping, picnicking, hiking, backpacking, hunting, fishing, boating, nature study.

Wildlife: Endangered or threatened bald eagle and American peregrine falcon.

THE ELEVATIONS in the Prescott National Forest sharply range from 3,000 feet to 8,000 feet, with the result that visitors have the opportunity to sample several distinct vegetation patterns.

The lowest elevation is at the eastern edge in the vicinity of Camp Verde. It is desert country, and you will find arid-loving plants thriving—plants such as yuccas, agaves, and many kinds of cacti. As elevations increase upward to 5,000 feet, you can see dense growths of evergreen shrubs, called chaparral. One good spot to sample them is around Tonto Springs west of Prescott.

Much of the forest between 5,000 and 7,000 feet, particularly north of Prescott and Jerome, is made up of woodlands where pinyon pine, juniper, and several kinds of oak grow. The Lonesome Pocket Trail passes through these woodlands where it begins at Henderson Flat. The trail then climbs 1,500 feet in less than two miles in sandstone country. The final half mile is extremely steep as it reaches the Mogollon Rim.

On the highest reaches, in the 6,000-to-8,000-foot range, are stands of ponderosa pines. Spruce Mountain and Mt. Union, both areas south of Prescott and reached via Senator Road, are excellent examples of this evergreen forest. There are small and welcome picnic areas at both mountains.

Parts of the Sycamore Canyon and Pine Mountain wildernesses are in the Prescott and they are particular areas of attraction. Sycamore Canyon, which extends into the Kaibab and Coconino forests, can be reached by taking the

Packard Trail (#66) near the mouth of Sycamore Creek and the Verde River. After about a mile and a half of steep climbing, the trail connects with the easier Sycamore Trail, and you will begin to get panoramic views of the colorful canyon. The Sycamore Trail continues for several miles.

The 20,062-acre Pine Mountain Wilderness is shared with the Tonto National Forest. The dominating feature is the Verde Rim, which also serves as the dividing line between the Prescott and the Tonto. The rim offers sweeping views of this rugged desert country. Again, the terrain is exceptionally steep, and there are many deep canyons.

Finally, for a touch of nostalgia, proceed south from Prescott, past Spruce Mountain, to Palace Station, where one of the few remaining stagecoach stations from a lost era is located. The log station structure dates back to 1875 and contains two downstairs rooms where food and drink were most likely dispensed, a kitchen, and a sleeping loft. It originally served as a halfway station between Prescott and the active Peak Mine.

Tonto National Forest

Location: 2,873,917 acres in south-central Arizona northeast of Phoenix. U.S. Highway 77 and State Routes 69, 87, and 88 are the major highways. Maps and brochures available at the district ranger stations in Carefree, Globe, Mesa, Payson, Tonto Basin, and Young or by writing the Forest Supervisor, 102 S. 28th St., Phoenix, AZ 85038.

Facilities and Services: Campgrounds (some with camper disposal service); picnic areas with grills; boat launching areas; swimming beaches.

Special Attractions: Desert habitats; Arizona "diamonds"; four wilderness areas.

Activities: Camping, picnicking, hiking, backpacking, hunting, fishing, boating, swimming, nature study, horseback riding.

Wildlife: Endangered or threatened bald eagle and American peregrine falcon; desert gopher tortoise.

THE SETTING of Zane Grey's *Under the Tonto Rim* is beneath the Mogollon Rim in what is now the Tonto National Forest. Grey, in fact, lived "under the rim" for a while, and his cabin can be reached by a short hike from Forest Road 289 north of Kohl's Ranch.

The Tonto is a land of contrasts, from dry desert where saguaros, cholla, and other cacti grow, up to 8,000-foot mountains with firs, spruces, and ponderosa pines. Rocky Tonto Creek, at Kohl's Ranch, is one of the finest trout streams in the forest. West of Kohl's Ranch is fascinating Diamond Point, a forest prime attraction. It is a mountain with an abundance of sparkling quartz crystals, known as Arizona diamonds. Needless to say, many visitors enjoy the dazzling trip. From the summit of Diamond Point, overlooking giant century plants, is a superb 360-degree vista.

A good way to start exploring the forest from Diamond Point is to follow the Beehive Highway (Route 87) to Phoenix, skirting the eastern edge of the Mazatzal Wilderness. One mile south of the road to Tonto Basin, the first saguaro appears. The hills in this area are covered with century plants, yuccas, prickly pears, and junipers. Along the east side of Route 87, the Forest Service has a major exhibit known as Desert Vista. To the east you can see the silhouetting spire of so-called Weaver's Needle in the Superstition Mountains, although it is more than 20 miles away. Between the vista and the skyline, as far as the eye can see, is the desert, with saguaros, barrel cacti, chollas, prickly pears, ocotillos, and paloverdes. For those with keen interest in desert flora, there is a trail with signposts that loops a short way into the desert.

East of Phoenix is Highway 88, the Apache Trail, a worthwhile scenic route that provides access to Canyon Lake, Apache Lake, and Roosevelt Reservoir as it proceeds from Apache Junction to the Miami and Globe area. The road is paved at both ends, but there is a long stretch of rough gravel in between. The lakes are extremely popular for boating, swimming, and fishing, and there are pleasant campgrounds near each. South of Miami are the wild and woolly Pinal Mountains. The Sixshooter and Upper Pinal campgrounds, at 7,500 feet, are delightfully cool at night and make you feel like one of the pioneers of old. To reach the Pinals, take Forest Road 580 south from Miami and across Madera Peak.

In addition, there are several wilderness areas in the Tonto that are ideal for exploring the backcountry, and all are special forest attractions. One is the Sierra Ancha Wilderness, nearly 21,000 acres of high cliffs, box canyons, and small mesas west of Cherry Creek. The area contains several fascinating Indian ruins, including cliff dwellings. By hiking up Devil's Chasm in the wilderness, you can travel through and observe semidesert to chaparral to forest growth. The semidesert area near Cherry Creek contains the green-stemmed paloverde, ironwood, mesquite, and many spine-bearing plants such as saguaro, buckhorn cholla, catclaw, prickly pear cacti, and ocotillo. In the chaparral the vegetation is dominated by mountain mahogany, manzanita, Apache plume, and the holly-leaf buckthorn. At higher elevations are white fir, Douglas fir, and bigtooth maple.

Another wilderness area is the Mazatzal which contains 205,000 acres at the north end of the Mazatzal Mountains. There are many steep, V-shaped canyons between the high and precipitous ridges. The most undisturbed region lies along its eastern edge between Mazatzal and North peaks. There are spectacular scenery, Indian ruins, and vegetation that ranges from desert plants to chaparral to forests of ponderosa pine and Douglas fir.

Most famous of the Tonto wilderness is the Superstition Wilderness, partly because of its proximity to Phoenix and partly because of the fascination of the area. There are many stories about the Superstition, including the one that the Lost Dutchman Gold Mine, reputed to be one of the potentially largest gold mines in the United States, is located in the mountain range. The trails are easier here, including the one from the Dons Camp up Peralta Canyon to Weaver's Needle. Although Weaver's Needle is the most prominent feature, there are other visitor attractions, such as Battleship Mountain, Iron Mountain, Hieroglyphic Spring, Fraser Canyon, and more Apache Indian dwellings along the cliffs of Rogers Canyon. Much of the area is composed of desert scrub vegetation.

Carson National Forest

Location: 1,491,000 acres in northern New Mexico, surrounding Taos. U.S. Highways 64, 84, and 285 and State Routes 3, 111, and 230 are the important accesses. Maps and brochures available at the district ranger stations in Blanco, Canijilon, El Rito, Penasco, Questa, Taos, and Tres Piedras, at the Ghost Ranch Visitor's Center at Abiquiu, or by writing the Forest Supervisor, Forest Service Building, Box 558, Taos, NM 87571.

Facilities and Services: Campgrounds (some with camper disposal service); picnic areas with grills; visitor's center and museum northwest of Abiquiu; ski facilities.

Special Attractions: Wheeler Peak Wilderness; Pecos Wilderness; Latir Peak Wilderness; Cruces Basin Wilderness; Chama River Canyon Wilderness; Wild and Scenic rivers.

Activities: Camping, hiking, backpacking, hunting, fishing, nature study, skiing, horseback riding, river rafting.

Wildlife: Endangered or threatened bald eagle and American peregrine falcon.

WHEN YOU HIKE in the mountains around Taos, New Mexico, you have a keen awareness that relicts from ancient Indian cultures and early Spanish settlers are all around you. As you drive the many scenic forest roads in the Carson, you will see some of this history and culture at such communities as Taos, Truchas, Las Trampas, and Cordova.

The Carson has much to offer. For one thing, Wheeler Peak, the highest mountain in New Mexico at 13,160 feet, a few miles northeast of Taos, is in the heart of the Wheeler Peak Wilderness. It can only be reached by hiking or horseback. The beauty of this high country wilderness (most of the 6,029 acres are above timberline) is hard to match. Rocky peaks and ridges, some with steep slopes of loose rocks, others with chutes where avalanches have gone before, are abundant. Deep blue high mountain lakes, such as Bear Lake and Lost Lake, all filled with native trout, are waiting for your camera or your fishing line. Restful open meadows, with a myriad of colorful wildflowers, contrast with the dense green forests of lower elevations. Near the summit of Wheeler Peak, the barren-looking areas are in actuality covered by the mat-forming alpine fescue grass, several sedges, and the golden avens of the rose family. Be alert for the showy Hayden painted cup (a type of Indian paintbrush), the snowy erysimum of the mustard family, and the lavender-blue bellflower. The only mammals apt to be seen are the scurrying, mouselike pike or coney, the Colorado chipmunk, and the yellow-bellied marmot.

Four other wildernesses, the Pecos, Latir Peak, Cruces Basin, and Chama River Canyon, are in the Carson. To get to the Pecos, follow State Route 73 from Pernasco to its end at the Santa Barbara Campground. From here follow the trail along the West Fork of the Rio Santa Barbara south for about a mile. Or take the Las Trampas Canyon Road to the Trampas Canyon Campground for a hike to lofty Jicarilla Peak and its surprise Hidden Lake.

The 12,708-foot Latir Peak, the main attraction of the Latir Peak Wilderness, is best approached by a ten-mile circuitous trail from Cabresto Lake Campground. The trail follows rippling Lake Fork and Bull Creek before swinging north past Virgin Canyon and Vendo Peak. The Cruces Basin Wilderness, nestled between Brazos Ridge and Toltec Mesa, extends nearly to the Colorado border. Only 2,900 acres of the Chama River Canyon Wilderness is in the Carson. The trail to it is located a few miles west of the Ghost Ranch Visitor's Center at Abiquiu. It takes its name from the picturesque gorge formed by the Rio Chama.

Another attraction of the Carson is its rivers. The Rio Grande and the Red are two in the Carson that have been designated Wild and Scenic rivers. Both offer gouged gorges along which you can hike or down which you can run the rapids. The Rio Grande has some dangerous stretches of rapids, and amateurs should be aware of this. Fishing for melt-in-your-mouth rainbow and brown trout is excellent here.

If you like secluded campsites and do not mind getting to them over rough roads, there are several we recommend. Cabresto Lake, at 9,200 feet, is our favorite, partly because fishing is good in the lovely blue lake here and partly because from here you can hike to the Latir Peak Wilderness.

The isolated campgrounds near the Canjilon Lakes and Trout Lakes are also ideal for a serene environment.

Sometime during your visit to the forest, stop at the Ghost Ranch Visitor's Center 14 miles northwest of Abiquiu, where beavers are the featured subject. There is a miniature, man-made "Beaver National Forest" here, as well as exhibits on native plants and animals (including the beaver), geology, and fossils. Three miles north of the visitor's center is the so-called Echo Amphitheater, a natural wonder carved into a sheer cliff wall and from which echoes resound. A camp-ground is nearby.

For winter sports enthusiasts, there are ski areas at Sipapu, Twining, and Red River.

Cibola National Forest

Location: 1,637,000 acres in eight units in central and western New Mexico, mostly in the vicinity of Albuquerque and westward. Interstates 40 and 25, U.S. Highways 54, 60, and 85, and State Routes 53 and 78 are the major highways. Maps and brochures available at the district ranger stations in Clayton, Magdalena, Mountainair, Grants, and Tijeras or by writing the Forest Supervisor, 10308 Candelaria NE, Albuquerque, NM 87112.

Facilities and Services: Campgrounds (some with camper disposal service); picnic areas with grills; visitor's center at Sandia Crest.

Special Attractions: High mountains with unsurpassed views; unusual stand of Rocky Mountain maple; four wilderness areas.

Activities: Camping, picnicking, hiking, backpacking, hunting, limited fishing, horseback riding, nature study.

Wildlife: Rocky Mountain bighorn sheep, mountain lion, golden eagle, turkey vulture, black bear, gray fox, coyote, Colorado chipmunk, mule deer, elk, antelope, and numerous birds.

FROM THE Cibola National Forest observation deck on Sandia Crest at an elevation of 10,678 feet, you have an uninhibited and unsurpassed view of New Mexico in every direction, a view that encompasses about 12,000 square miles. To the west you can see 11,368-foot Mount Taylor and the Zuni Mountains, approximately 65 and 95 miles away, respectively; to the southwest are the Magdalena and San Mateo Mountains, 95 miles away; to the southeast are the Gallinas Mountains, 77 miles distant. To get an idea of the vastness of the Cibola, consider that all of these mountains are a part of it.

We suggest starting at the north end of Sandia Crest, with its breathtaking view of Albuquerque at its western foot. Sandia Crest can be reached by a privately operated tramway, by hiking, or by driving up the eastern face of the mountain on State Routes 44 and 536. No matter how you get there, you will climb more than 4,000 feet and pass through four biological life zones. These vegetation patterns are the same you would encounter if you drove from Albuquerque to Canada.

At about 6,000 feet near the ranger station until you get to an elevation of 7,200 feet, you are in the pinyon-juniper life zone. The soils that support the pinyon pines and western junipers, you will see, are shallow and rocky. If you pass through this zone in early morning or at dusk, you might be lucky and catch a glimpse of mule deer, black bears, porcupines, spotted skunks, Colorado chipmunks, long-tailed weasels, coyotes, or gray foxes. Frequently overhead are turkey vultures and sometimes a mightly golden eagle.

From 7,200 to 7,800 feet is an area dominated by ponderosa pines, although the composition of the vegetation will depend on whether it is a moister north-facing slope or a more arid south-facing slope. The cooler north face is likely to have white fir and Douglas fir on it. As for animal life in the ponderosa pine zone, in addition to mule deer, bear, and porcupine, look for the Abert squirrel and the ring-tailed cat, and listen for the calls of a western wood pewee, a warbling vireo, and a mountain chickadee.

At about 7,800 feet, oaks and aspens mix with ponderosa pines and firs. Numerous birds live in this zone; we were enthralled by the red crossbill, the pygmy nuthatch, Clark's nutcracker, the green-tailed towhee, the orange-crowned warbler, and many others. If you try to picnic, loudly chattering jays will probably join you.

Above 9,800 feet is the cold spruce-fir zone, where Engelmann spruce, subalpine fir, and Douglas fir live in some of the cleanest and freshest air you will ever breathe. This is the zone for the remarkable Rocky Mountain bighorn sheep, recognized by its great curved horns, and the elusive mountain lion. Birds that make it up this far include the northern three-toed woodpecker, the red-breasted nuthatch, and the white-throated swift.

Several interesting hiking trails are recommended in the Sandia Mountains. It is 10 miles by trail from Sandia Crest north to Tunnel Springs, where the natural spring water provides suitable habitat for an oasis of western cottonwoods and box elders. Along the trail is a view overlooking Piedras Lisa Canyon. Another worthwhile hiking trail is the La Luz. You can learn much about the geology of the area from some of the rock ledges along the way, and wildflower enthusiasts will thrill at the site of the Rocky Mountain iris, purple-spiked verbena, the dainty Franciscan bluebell, scarlet penstemon, columbine, and the showy mariposa lily. For the handicapped, the Cienega Nature Trail into a wet meadow on the eastern slopes is recommended. A part of the Sandia Mountains is included in the Sandia Mountain Wilderness, one of the major forest sights.

In the eastern part of the forest, a few miles south of the Sandias and southeast of Albuquerque, is another part worth a visit. It is the narrow range of Manzano Mountains, the home of the largest stand of the uncommon Rocky Mountain maple. Related to the sugar maples of the east, this mountain maple displays equally vivid colors of red, orange, and yellow during late October. We recommend going to the Fourth of July Campground, eight miles west of Tajique. At 7,400 feet, this is an area where you can camp or picnic beneath tall, swaying pines. There is a trail leading up a narrow, damp canyon lined with Rocky Mountain maples. If you are energetic, continue up the adjoining Tajique Canyon Trail for two miles and join the crest trail in the Manzano Mountain Wilderness, another popular attraction of the forest.

In the western part of Cibola, on either side of Grants, New Mexico, are two more popular segments of the forest. One area is dominated by 11,398-foot Mt. Taylor, an old dormant volcano that is the highest elevation in the Cibola. A trip to the summit is a challenging and exciting experience. By vehicle you can follow paved Route 547 up Lobo Canyon for about 15 miles; then turn right onto a gravel forest road. Along the route you will see large groves of aspen mixed with ponderosa pine. Chunks of lava remind us that Mt. Taylor was once an active volcano. You will have to stop at the La Mosca Lookout Tower and climb the last 2 miles to the summit on foot, but it is well worth the effort. Since the altitude is high, be sure and pace yourself. You are in elk territory, as well as bobcat, bear, coyote, and mountain lion land. A full circle view and a nearly pure stand of Engelmann spruce are your rewards for making the climb.

West of Grants are the 9,000-foot Zuni Mountains. This is also a lava land reminiscent of lunar landscapes. In addition, there are craggy bluffs and rocky spires dominated by pinyon and juniper trees at lower elevations and ponderosa pine and Douglas fir higher up. The most popular recreation site is at McGaffey Lake south of Gallup, one of the few places in the Cibola suitable for fishing.

Two rugged and less visited mountains must be mentioned. One is the Gallinas, an isolated range in central New Mexico lying west of Corona and some 80 miles southeast of Albuquerque. It is a must for the adventurer. Abandoned log houses are in the area, as well as the remains of an old Civil War army post that is a short distance south of Gallinas Spring. Forest Road 99 up Red Cloud Canyon past Gallinas Spring to the Gallinas Lookout near 8,637-foot Gallinas Peak takes you there. While hiking, be alert for the interesting band-tailed pigeon which whoo-hoos like an owl. If you want to linger, there is an adequate campground called Red Cloud at an elevation of 7,600 feet.

The other less visited mountains are the Magdalena, San Mateo, and Datil located west of Socorro in the vicinity of Magdalena. To get a feel for these ranges, make camp near the quiet stream in the Water Canyon Campground where narrow-leaved cottonwoods form intense shade and the antics of the acorn woodpeckers are most entertaining. After a strenuous hike of several miles along the South Baldy Trail, through pinyons and junipers, ponderosa pines and aspens, and eventually Douglas firs, you will arrive at the towering 10,700-foot South Baldy Peak. Two wilderness areas, the 19,000-acre Withington Wilderness and the

40,000-acre Apache Kid Wilderness, are in the steep, rugged, rocky San Mateo Mountains south of the Magdalenas. Chances are you will encounter elk, black bear, mule deer, antelope, and perhaps a mountain lion in these areas.

Gila National Forest

Location: 3,320,135 acres in western New Mexico north of Silver City. U.S. Highways 60 and 189 and State Routes 61 and 78 are the major accesses. Maps and brochures available at the district ranger stations in Truth or Consequences, Glenwood, Luna, Mimbres, Quemado, Reserve, and Silver City, at the Gila Cliff Dwellings Visitor's Center, or by writing the Forest Supervisor, 2610 N. Silver St., Silver City, NM 88061.

Facilities and Services: Campgrounds (some with camper disposal service); picnic areas with grills; visitor's center; boat launching ramps.

Special Attractions: Unusual Catwalk; Gila Wilderness; Aldo Leopold Wilderness; Blue Range Wilderness.

Activities: Camping, picnicking, hiking, backpacking, boating, hunting, fishing, horseback riding.

Wildlife: Endangered or threatened bald eagle, American peregrine falcon, Gila trout, and Gila topminnow; rare banded rock rattlesnake, spotted bat, hooded skunk.

WE BEGAN our exploration of the Gila at its fascinating Catwalk on the western edge of the forest. The Catwalk is a narrow and sometimes dramatic trail that follows the massive rock walls of Whitewater Canyon. The walk parallels the three-mile route of the water pipe used at the turn of the century to carry water from mines down to the little village of Graham and its mill. At a later date a larger pipeline was constructed over the stream, part of which can be seen today. At places along the canyon where the sheer walls fall to the rushing stream's edge, the trail follows a wooden walkway attached to the cliff and suspended above the water. The trail through the cool and colorful canyon is not difficult. Camping and picnicking facilities are adjacent.

From the Catwalk area, we suggest driving the breathtakingly scenic Bursum Road through some of the Gila high country and near the vast Gila Wilderness. Leave U.S. Highway 180 on State Route 78 just south of Alma and proceed on Bursum Road. In a few miles you come upon the nearly abandoned mining village of Mogollon with its Fannie Hill Mine perched precariously on the mountain above town. Beyond Mogollon the highway follows a tortuous route above Silver Creek and eventually across Whitetail Canyon until it comes to the Bursum Campground and Picnic Area, at 8,500 feet the highest developed recreation site in the forest. It is less than one mile to the northern edge of the Gila Wilderness and only a few miles via trail to 10,892-foot Whitewater Baldy Peak. The hike is strenuous, but it passes through some great woods of ponderosa pine to forests of Engelmann spruce, white fir, cork-bark fir, Douglas fir, limber pine, and aspen, to a treeless tundra with low, matted vegetation. Be prepared for sudden summer thunderstorms, and stay off the ridges during lightning.

From Silver City there is access to the fabulous Indian cliff dwellings found in the forest. We took State Route 15 north following Copperas Creek between two lobes of the Gila Wilderness until the road dead-ended at the Gila Cliff Dwellings. The dwellings are a designated National Monument; the National Park Service and the forest jointly operate a visitor's center here. You must, however, endure a half-mile, semirugged trail to and through the cliff dwellings. The dwellings themselves are about 180 feet above the canyon floor. There is often a ranger near the dwellings to explain the structures.

The nearby Gila Wilderness is something special and a major forest attraction. It was the first designated National Wilderness Area; it was established in 1924 after Aldo Leopold, a forest service employee, pointed out the fragile and rare ecology of the wild land. The 569,600-acre region, the largest roadless area in New Mexico, is dominated by the sprawling Mogollon Mountains, and is known for its one-of-a-kind rock spires and pinnacles. In addition, the largest stand of virgin, never-to-have-been-cut ponderosa pine is found here. Further, there is a giant network of trails, although most are long and strenuous and require good physical condition on the part of the hiker or the horseback rider, but you will see both Rocky Mountain mule deer and white-tail deer and even possibly an antelope and javelina. The shy mountain lions are present, and so are Rocky Mountain bighorn sheep and elk that have been introduced from other areas of the West. From time to time the threatened bald eagle and American peregrine falcon can be seen overhead. What is more, in some of the streams are the endangered Gila trout and Gila topminnow. Other rare animals indigenous to the wilderness are the banded rock rattlesnake; spotted bat, whose high-pitched clicking may puzzle you as you try to sleep in your primitive campsite; hooded skunk; and uncommon birds such as the osprey, golden eagle, goshawk, and elf owl. You will probably come upon ruins of old cabins and an occasional grave marker. An interesting way of penetrating the wilderness during the spring runoff is to float on rafts down the Gila River. It is best to put in below Gila Hot Springs along Route 15. (You will eventually emerge from the wilderness northeast of the community of Gila.)

The Aldo Leopold Wilderness, named for the man who brought attention to the wild state of the Gila, is east of the Gila Wilderness. It and the Blue Range Wilderness west of the San Francisco River offer additional backcountry opportunities and are special forest attractions.

For less rugged activities, a stay at scenic Lake Roberts along State Route 35 off the southern edge of the Gila Wilderness is recommended. You can fish, boat, hike a couple of short nature trails, or do nothing but relax.

Lincoln National Forest

Location: 1,100,000 acres in southern New Mexico, mostly between Alamogordo, Carrizozo, and Roswell. U.S. Highways 54, 70, 82, and 380 and State Routes 24 and 137 pass through or near the forest. Maps and brochures available at the district ranger stations in Cloudcroft, Carlsbad, Mayhill, and Ruidoso or by writing the Forest Supervisor, 11th and New York, Alamogordo, NM 88310.

Facilities and Services: Campgrounds (some with camper disposal service); picnic areas with grills; museum.

Special Attractions: Smokey Bear Museum; White Mountain Wilderness; Capitan Mountain Wilderness; limestone caves.

Activities: Camping, picnicking, hiking, backpacking, hunting, limited fishing, nature study, skiing, horseback riding, spelunking.

Wildlife: Endangered or threatened bald eagle and American peregrine falcon.

THE LINCOLN is known as the birthplace of Smokey Bear, for it was in this forest that the cub who became internationally known was rescued after a devastating forest fire in 1950. The Smokey Bear Museum near Capitan is dedicated to him and contains exhibits of the forest.

One of the great wilderness areas in the United States is the Lincoln's White Mountain Wilderness northwest of Ruidoso. It is a popular forest attraction. This 31,283-acre area is highlighted by extreme elevational differences— from 6,500 feet to 11,260 feet. Life zones range from the Lower Sonoran Desert type in which you find the highest heat tolerance of desert life to the frigid Alpine Fir type at Lookout Mountain.

Another forest attraction is the Capitan Mountain Wilderness northeast of Capitan. Those who enjoy high altitude hiking will particularly like it. There is Summit Trail connecting 10,083-foot Capitan Peak to 10,179-foot Summit by way of Pierce Canyon Pass. Lateral trails into Pierce Canyon and Seven Cabins Canyon are also rewarding for their scenery.

Although not a designated wilderness area, the southern part of the Lincoln which extends to the Texas border and includes a part of the Guadalupe Mountains is as wild as anyplace. This area is bounded on the east by Carlsbad Caverns National Park and on the south by Guadalupe Mountain National Park, and the terrain shares characteristics of both. If you seek solitude, this is the place to go. There are no developed campgrounds, although there is a picnic area near Sitting Bull Falls, named after the famous Indian chieftain. This region is extremely rugged and contains many limestone caves and several deep gorges worth exploring. Cottonwood Cave is open to the public, but the back part cannot be entered unless a forest ranger accompanies you. Inquire at the ranger station in Carlsbad about other caves. At elevations below 5,000 feet, the vegetation is typical of the Chihuahuan Desert and includes the mescal agave, lechuguilla, sotol, yucca, and creosote bush. Between 5,000 and 6,500 feet, the plant life consists of pinyon pine, alligator juniper, one-seeded juniper, mountain mahogany, and prickly pear cactus.

Among the outstanding hiking trails is the 2½-mile loop Osha Trail, which offers views of the Tularosa Basin and the White Sands National Monument where great quantities of gypsum form towering white dunes. This trail begins opposite a railroad trestle on Highway 82. More strenuous is the 13½-mile Rim Trail which follows a crest in the Sacramento Mountains; it, too, offers grand views of the Tularosa Basin. For a short hike, the half-mile trail to Nelson Vista overlooking the White Sands is recommended. There is a special trail for the handicapped that originates at the Upper Sleepy Grass Picnic Area.

We sampled some of the trout fishing and found it excellent in Bonito Lake, Nogal Lake, and a few of the mountain streams. We also enjoyed lunch at different picnic areas, a beautiful one being at Monjeau at an elevation of 10,000 feet.

Equally well known is the Sierra Blanca Winter Sports Area, with its multispired lodge at 12,000 feet, one of the southernmost ski areas in the United States. You can enjoy Sierra Blanca in the summer, as well, by hiking the trails, riding horseback, or just plain taking it easy in the fresh mountain air.

Santa Fe National Forest

Location: 1,589,000 acres in northern New Mexico, on either side of Santa Fe. U.S. Highways 84, 85, and 285 and State Routes 4, 63, 96, and 126 are the major highways. Maps and brochures available at the district ranger stations in Coyote, Cuba, Espanola, Jemex Springs, Las Vegas, Pecos, and Santa Fe or by writing the Forest Supervisor, 1220 St. Francis Drive, Santa Fe, NM 87501.

Facilities and Services: Campgrounds (some with camper disposal service); picnic areas with grills.

Special Attractions: Pecos Wilderness; San Pedro Parks Wilderness; Chama River Gorge Wilderness; Dome Wilderness.

Activities: Camping, picnicking, hiking, backpacking, hunting, fishing, winter sports.

Wildlife: Endangered or threatened bald eagle and American peregrine falcon; mule deer, bighorn sheep, elk, black bear, Abert squirrel.

NEW MEXICO SCENERY is at its finest in the Santa Fe National Forest, from the lofty Sangre de Cristo Mountains and its 13,103-foot Truchas Peak on the east to the Jemez Mountains and its nearly 12,000-foot Chicomo Peak on the west. In between, the land masses drop sharply down to the level of the Rio Grande River which divides the forest into its two major units.

Imagine crawling into your winter sleeping bag in the middle of July! You will be happy to do so if you decide to camp at Santa Fe Basin, the highest campground in the forest at 10,300 feet. From here it is less than a mile to the Pecos Wilderness and a series of trails that lead to Santa Fe Baldy far above timberline and scenic Spirit Lake. The Pecos Wilderness is a big crowd-drawer as it is nationally known for its spectacular, postcard-pretty scenery: rugged peaks, evergreen- and aspen-clad forests, and clear mountain lakes. Particularly attractive are the thundering Pecos Falls of the Pecos River and the Truchas Lakes. The trails are through woods dominated by ponderosa pine, Engelmann spruce, Douglas fir, cork-bark fir, white fir, limber pine, bristlecone pine and, of course, golden aspen. The animals most likely to be glimpsed are mule deer, bighorn sheep, elk, black bear, and the Abert squirrel. If you have room for your fishing rod, take it along, because the lakes and streams have a tantalizing offering of trout—rainbow, brown, cutthroat, and golden.

By contrast to the famous Pecos, the 41,000 acres of the San Pedro Parks Wilderness, another forest must, are topographically different. Despite an elevation that reaches

12,000 feet, this is an area of rolling gentle hills and large, open meadows filled with spring and summer wildflowers. For a taste, we recommend a drive to the end of rough Forest Road 14 and a hike along Trail #30. In only three miles you can go from the Rio Gallina through dense evergreen forests and pure stands of aspen to the forest's Red Rock Cliffs. From here either turn back or continue on and join a network of trails in the wilderness.

Another wilderness of note is the Chama River Gorge Wilderness which encompasses the narrow and winding gorges of the Chama and Gallina rivers. A few pronghorn antelopes live in this region.

If there is time, go on to the smaller Dome Wilderness which features the conspicuous 8,463-foot St. Peter's Dome, so-called because it has a fanciful resemblance to the dome of St. Peter's Church in Rome. Capulin Canyon is also located in this region.

For a pleasant drive, particularly when the aspens turn a brilliant golden hue in October, take State Route 4 north through Jemez Canyon to the rushing Jemez Falls. On the way, it is a great experience to picnic at Battleship Rock, a huge vertical monolith that rises dramatically above the surrounding pines as solidly as a battleship and takes you back aeons in time.

PACIFIC NORTHWEST

The forests of the Pacific Northwest total 21; those discussed are the Deschutes, Fremont, Malheur, Mt. Hood, Ochoco, Rogue River, Siskiyou, Siuslaw, Umatilla, Umpqua, Wallowa-Whitman, Willamette, and Winema in Oregon; Colville, Gifford Pinchot, Mt. Baker–Snoqualmie, Okanogan, Olympic, and Wenatchee in Washington; and the Chugach and Tongass in Alaska.

Deschutes National Forest

Location: 1,600,613 acres in central Oregon around Bend. U.S. Highways 20, 97, and 126 and State Routes 31 and 58 are the major roads. Maps and brochures available at the district ranger stations in Bend, Crescent, and Sisters, at the Lava Lands Visitor's Center, or by writing the Forest Supervisor, 211 NE Revere Ave., Bend, OR 97701.

Facilities and Services: Campgrounds (some with camper disposal service); picnic areas with grills; boat ramps; swimming beaches; visitor's center.

Special Attractions: Lava casts and other volcanic phenomena; four wilderness areas.

Activities: Camping, picnicking, hiking, backpacking, hunting, fishing, boating, swimming, horseback riding, winter sports, nature study.

Wildlife: Bald eagle, Rocky Mountain bighorn sheep, mountain goat, mountain lion, bobcat, moose, elk, osprey.

SINCE MANY of the special attractions of the Deschutes Forest are centered around volcanic activity, we recommend you start your exploration of the forest at the Lava Lands Visitor's Center south of Bend. The visitor's center, a modernistic building set at the foot of Lava Butte, provides a wealth of information about the geology of the area. Well-prepared exhibits, including an animated one, are worth studying. Proceed from the visitor's center to the summit of Lava Butte, either on foot or by a narrow road that circles the butte as it climbs. Lava Butte is a 500-foot-tall cinder cone. From the summit you can look down into the crater, 150 feet deep. On a clear day you can see the formidable Three Sisters Mountains and Broken Top to the west, Mt. Shasta on the southwestern horizon, and Newberry Crater just a short distance south.

Your next stop is across Route 97 at Lava River Cave. A forest ranger will rent you a lantern and show you the entrance to the cave, in actuality a subterranean lava tube that extends for several hundred yards. You may wish to explore the entire length of the tube, but we decided we had seen all there was after about 300 feet. There are none of the usual cave formations, since this is not a limestone cave with delicate formations.

Proceed from the Lava River Cave to the Lava Cast Forest, truly a remarkable phenomenon. Pick up a leaflet for your self-guided trail and begin the circular, mile-long trail. You will soon see the star features of the forest, the lava casts. During volcanic activity long ago, slow flowing lava moved down the slopes of Mt. Newberry and into a forest where the lava chilled and formed a protective coating around each tree. Through the years the charred wood inside this lava coating rotted away, leaving only the lava casts. Some casts protrude above ground, others are sunken in the ground; still others lie horizontally. This is the most extensive lava cast forest in the world.

Next stop is Newberry Crater, reached by an all-weather road to the summit. Ten thousand years ago, massive Mt. Newberry began to spring leaks through which lava flowed. As the lava drained from the mountain, the top of the volcano collapsed, resulting in a central depression nearly five miles across. Within the depression Paulina and East lakes were formed, and the wildly gushing Paulina Creek Falls developed. The results of a great obsidian, or glass-like rock, flow are evident in the form of massive black volcanic glass just a short distance southeast of Paulina Lake.

In the area surrounding Mt. Newberry are a number of other caves, some containing perpetual masses of ice. One unique cave, visited only when accompanied by a ranger, is Lavacicle Cave. Hanging from its ceiling are black, solid lava stalactites. One other feature that fascinated us was Hole-in-the-Ground, a vast volcanic explosion crater that is a half mile across from rim to rim.

By contrast, that part of the forest that lies west of Bend is a land of wild and woolly high mountains. Bachelor Butte is most accessible and is, in fact, overcrowded because it is a year-round ski resort. When we were there during the third week in June, we were hit with a snowfall. West of the ski lodge but before reaching Sparks Lake are some red-stained Indian pictographs on large boulders visible from the road.

The lands west of Bend that surround Mt. Jefferson, Mt. Washington, the Three Sisters, and Diamond Peak are wild and roadless and are designated as four different wilderness areas. All are prime attractions and are shared with the Willamette National Forest; Mt. Jefferson is also partly in the Mt. Hood National Forest.

The Diamond Peak Wilderness boasts of 35,440 acres, with its Diamond Peak at 8,744 feet. The peak is popular with climbers. The Skyline Trail, completely within the forest and 13 miles long, connects Odell and Crescent lakes. It is a relatively easy trail. Other features of this wilderness are discussed under the Willamette Forest entry.

Although the larger part of the 196,708-acre Three Sisters Wilderness is in the Willamette National Forest, the three mountain peaks from which the wilderness takes its name straddle the line between the Deschutes and the Willamette. Broken Top Mountain, a conspicuous landmark with Bond and Crook glaciers hanging on its eastern face, is entirely in the Deschutes, as are the three beautiful Green Lakes. There is a good trail to these lakes that follows Fall Creek and skirts the edge of two extensive lava flows.

Steep-faced Mt. Washington, with deeply grooved, glacial-carved ridges, is the central feature of the Mt. Washington Wilderness, a 46,655-acre roadless area that lies mostly within the Willamette National Forest. Most of the trails are in the Willamette and are discussed in that entry.

Although Mt. Jefferson, the prominent focal point of the Mt. Jefferson Wilderness, lies in the Willamette, there is still plenty to see in the Deschutes part of the wilderness. Three-fingered Jack Mountain, a favorite with experienced mountain climbers, raises its several pointed peaks skyward at the southern end of the wilderness. Do not try to climb it if you have never climbed a mountain before. Instead, take the 26½-mile Summit Trail from Hole-in-the-Wall Park to Square Lake, passing such sights as Table Lake, Patsy Lake, and Carl Lake. After leaving Carl Lake, there is a series of steep switchbacks to South Cinder Peak that may leave you breathless and with legs trembling. South of the peak the trail is routed near other blue lakes, including Rockpile, Wasco, Jack, and Booth, before coming to Square Lake, less than a mile from the trail's end along U.S. Route 20. For a shorter trip, take the 2-mile trail from the end of Forest Road 1210B near Abbot Butte to sparkling Cabot Lake.

In addition to all the volcanic phenomena southeast of Bend and the high mountain wilderness west of Bend, there are still a few other features you should try to work in to your visit. Pringle, Benham, and Dillon falls are three turbulent cataracts whose roars will drown out any conversation you may wish to engage in. By contrast, the headwaters of the Metolius River bubble from the base of Black Butte amid a picturesque setting of pines, firs, and spruces. It is a memorable sight.

We also recommend a trip over the Cascade Lakes Highway to Crane Prairie Reservoir where the forest service has preserved an osprey nesting area. Eaglelike in appearance, the uncommon osprey builds huge nests near the tops of trees. You can also compete with the osprey for rainbow and brook trout in the lake. Four comfortable campgrounds are situated around the edge of Crane Prairie Reservoir.

Fremont National Forest

Location: 1,198,308 acres in south-central Oregon east of Klamath Falls. U.S. Highway 395 and State Routes 31 and 66 cross the forest. Maps and brochures available at the district ranger stations in Bly, Lakeview, Paisley, and Silver Lake or by writing the Forest Supervisor, Box 551, Lakeview, OR 97630.

Facilities and Services: Campgrounds (some with camper disposal service); picnic areas with grills; boat ramps; winter sports facilities.

Special Attractions: Gearhart Mountain Wilderness.

Activities: Camping, picnicking, hiking, backpacking, hunting, fishing, horseback riding, boating, winter sports.

Wildlife: Black bear, bobcat, mountain lion.

IN THE HEART of the Fremont National Forest is the Gearhart Mountain Wilderness, 18,709 acres of rugged terrain with Gearhart Mountain the featured attraction. Although the mountain rises only to 8,364 feet, it is characterized by picturesque, jagged rock formations along the crest, giving the mountain a formidable appearance. Equally attractive on the mountain are the many wildflower meadows that are just below the ridgetops.

The best way to see all the wilderness is to hike the 12-mile Gearhart Trail which runs the entire length of it. The best place to begin is at the south end at Lookout Rock, just a short distance from Corral Creek Campground. The trail immediately comes to an area of rugged rocks and fascinating pillars, all carved by erosional forces. For ¾ of a mile, the trail parallels an impressive range of cliffs that rise to 400 feet above the surrounding terrain. Climaxing these cliffs is an isolated monolith known as the Dome. As you climb to Gearhart Mountain, you will be treated to many special sights in all directions. On the north side is Gearhart Notch, a sheer cliff that has broken away from the main part of the mountain. Leaving Gearhart Mountain, the trail drops steadily to a low, wet area known as Gearhart Marsh where typical wetland rushes and sedges prevail. From the marsh the trail proceeds east to Blue Lake, the only lake in the wilderness. It is then less than a mile to the northern boundary of the wilderness.

A feature that should be on your itinerary is the Abert Rim in the southeast corner of the forest. The rim, which rises 2,500 feet above the valley floor, is considered to be the largest exposed geologic fault in North America. A fault is a fracture in the earth's crust where the adjacent cliffs have slipped disproportionately away from each other. The top of the rim is covered by layers of lava.

Big Hole is an interesting volcanic phenomenon in the northern tip of the forest. It is a huge explosion crater, but because trees have filled the depression, it is difficult to perceive the vastness of the crater. This is in sharp contrast to nearby Hole-in-the-Ground in the Deschutes Forest, which is a similar explosion crater but does not have trees in it.

We stumbled across one of the most beautiful campgrounds in any national forest along Sprague River east of Bly. The fishing in the river is superb.

Malheur National Forest

Location: 1,460,000 acres in east-central Oregon, surrounding John Day. U.S. Highways 26 and 395 bisect the forest. Maps and brochures available at the district ranger stations in Burns, John Day, and Prairie City or by writing the Forest Supervisor, 139 NE Dayton St., John Day, OR 97845.

Facilities and Services: Campgrounds (some with camper disposal service); picnic areas with grills; swimming beaches; boat launching ramps.

Special Attractions: Strawberry Mountain Wilderness; Cedar Grove Botanical Area; Vinegar Hill–Indian Rock Scenic Area.

Activities: Camping, picnicking, hiking, backpacking, hunting, fishing, swimming, boating, horseback riding, nature study, winter sports.

Wildlife: Elk, black bear, bobcat, mountain lion, Rocky Mountain bighorn sheep.

IF YOUR FIRST APPROACH to the Malheur National Forest is by way of U.S. Highway 26, you will climb through dense coniferous forests to Dixie Summit, where there is a pleasant campground nestled beneath swaying pines. From Dixie Summit drive or hike the forest road about four miles north to Dixie Butte. From this high point is an outstanding view in all directions of the forest.

Wildest and most rugged in the Malheur is the 33,000-acre Strawberry Mountain Wilderness, with the broadly cone-shaped, 9,044-foot Strawberry Mountain, the highest peak, in the exact center of the wilderness. There are seven sky-blue lakes southeast of Strawberry Mountain, each with its own special features. We like High Lake which is nestled below a rock slide whose steep slope is capped by a rocky promontory known as Rabbit Ears. High Lake is reached by a mile-and-a-half trail from the end of Indian Springs Road. The high elevation trail that connects High Lake and Slide Lake passes by gnarly, white-trunked whitebark pine, some with only a small amount of living tissue present. Near the southwestern corner of the wilderness is Canyon Creek Natural Area, a beautiful protected stand of virgin ponderosa pine. Although ponderosas dominate this 700-acre tract, there are fine examples of Douglas fir, grand fir, and western larch present, as well.

Another botanical area listed as a forest attraction is Cedar

Grove, located along Cabin Creek in the Aldrich Mountains about two miles west of the Billy Fields Campground. Cedar Grove contains a 60-acre stand of Alaska cedar, a tree many miles from the next nearest colony of it.

The Malheur and Umatilla national forests share an area of rugged cliffs and wildflower-laden meadows in the Greenhorn Mountains that has been designated the Vinegar Hill–Indian Rock Scenic Area. It, too, is listed as a top forest attraction. Most of the area is above timberline, with Vinegar Hill at 8,131 feet and Indian Rock at 7,353 feet anchoring either end of the scenic area. Gravel roads lead to the summit of each of these mountains, where there are unending views into both national forests. The scenic area is in gold and silver territory, and the remnants of several old mines can be seen between Vinegar Hill and Indian Rock.

We were not disappointed in some of the unusual rock formations and fossil beds we had on our list to see. Arch Rock is a 15-foot-high opening in a rough-rock formation a short distance off of Forest Road 36, a few miles south of the ghost town of Susanville. A few miles farther south is Magone Lake and its weird slide area. A great landslide in the early 1800s formed Lake Magone and also exerted phenomenal pressure against the standing trees, causing them to grow at a severely tilted angle. A trail, whose surface is rough and irregular because of the landslide, passes beneath some of these tilted trees. The lake itself has a swimming beach and a boat ramp, and there is a well-developed campground along the shore. Rosebud Creek at the western edge of the Malheur is noted for its fossilized marine shells embedded in soft shale.

In the southeastern section of the forest, where mountains give way to rolling topography, the vegetation changes from coniferous forests to sagebrush and grasslands.

Mt. Hood National Forest

Location: 1,059,240 acres in northwestern Oregon, southeast of Portland. Interstate 80, U.S. Highways 26 and 30 and State Routes 35 and 224 are the major roads. Maps and brochures available at the district ranger stations in Dufur,

Maupin, Estacada, Troutdale, Parkdale, and Zigzag or by writing the Forest Supervisor, 2955 N.W. Division St., Gresham, OR 97030.

Facilities and Services: Campgrounds (some with camper disposal service); picnic areas with grills; boat ramps; swimming beaches; lodges; winter sport facilities.

Special Attractions: Timberline Lodge; Mt. Hood Wilderness; Multnomah Falls.

Activities: Camping, picnicking, hiking, backpacking, hunting, fishing, horseback riding, winter sports, nature study.

Wildlife: Black bear, mountain lion, bobcat.

A PAIR OF SAINT BERNARDS greeted us as we pulled in front of massive Timberline Lodge halfway up the south slope of mighty Mt. Hood. We had elected to stay in the lodge while we explored the Mt. Hood National Forest, and we were not disappointed. The wood and stone lodge, built as a project of the Works Progress Administration (WPA) in 1935, is now a National Historic Place. As we entered the lodge through the ski lobby, we were treated to a handsome carving of an Indian head on the heavy wooden door. One flight up is the main lobby and dining room, an architectural sight to behold. Newel posts at the head of the stairs have been made from old telephone poles and are intricately carved with Oregon wildlife. The hexagonal stone fireplace in the center of the main lobby rises 96 feet and has carved and inlaid wolves and mountain lions around it. Six huge pillars of ponderosa pine form the corner posts of this lobby. The twinkling light fixtures that hang from the ceiling are in the shape of Indian storage baskets. The iron gates into the elegant dining room have a fox-head design and snake handles. There is also a wooden door that weighs 1,000 pounds leading onto the stone balcony to test your strength! If the day is clear, you will see imposing Mt. Jefferson and the Three Sisters in the distance, with sparkling Trillium Lake at the base of Mt. Hood. Do plan to spend some time looking at the many relief carvings and oil paintings in the lodge.

The Magic Mile Chairlift and hiking trails to the top of Mt. Hood originate at the lodge. Although forest service brochures indicate that a beginning climber can reach the summit of Mt. Hood and back via the Hogsback Trail in 10 to 12 hours, the route is dangerous and requires the use of ice axes, rope, crampons, a compass, and plenty of stamina. The modern Wy'east Building across the parking area from Timberline Lodge is the new center for skiers.

Beautiful Mt. Hood, with its perpetual snow-white peak rising to 11,245 feet, is the central feature of the Mt. Hood Wilderness, a 47,100-acre high mountain area. As you might suspect, the wilderness, like the lodge, is a star attraction of the forest. A number of alpine meadows with bright-colored wildflowers dot the wilderness, and there are several gushing waterfalls.

The waterfalls along the Columbia River, however, are among the finest in the northwestern United States. The big forest drawing card, famous Multnomah Falls, is the most spectacular, falling for 542 feet before dropping a second time for 69 feet more. A short, popular trail leads from the base of the falls to a concrete bridge halfway up. A more rugged trail continues above and on to Larch Mountain five miles away. A short distance west of Multnomah Falls are the long and narrow cascades of Wahkeena Falls, while east of Multnomah are Oneonta, Horsetail, Elowah, and Tanner falls.

One of the most isolated areas in the forest is a 10,200-acre tract near the headwaters of the Clackamas River known as Bull of the Woods. There are at least 12 lakes set among high mountain scenery, and each is loaded with fish. Although the area is roadless today, there are signs of commercial activity long ago in the form of old mine shafts, tunnels, and rusty equipment.

Nearby and accessible only by a one-and-a-half-mile hiking trail is an area of natural hot springs known as Bagby Hot Springs. The largest spring flows 24 gallons of water per minute at a temperature of 138 degrees Fahrenheit. Only the old bathhouse, built in 1939 beneath towering firs, remains of the several buildings that were constructed around the springs. The area surrounding the hot springs is a plant lover's paradise, with dense stands of old western hemlock, Douglas fir, and western red cedar. Colorful thickets of rhododendron flower during late June and July.

Ochoco National Forest

Location: 950,000 acres in central Oregon north and east of Prineville. U.S. Highway 26 is the major access route. Maps and brochures available at the district ranger stations in Paulina, Prineville, and Burns or by writing the Forest Supervisor, Box 490, Prineville, OR 97754.

Facilities and Services: Campgrounds (some with camper disposal service); picnic areas with grills.

Special Attractions: Diversified rocks; Steins Pillar; Black Canyon Wilderness.

Activities: Camping, picnicking, backpacking, hunting, fishing, rock hounding, nature study.

Wildlife: Bald eagle, black bear, elk, bobcat, pronghorn antelope; wild horses.

THE OCHOCO is a forest for rock hounds and geology buffs. There are many well-known rock hound sites in the forest where a variety of colorful stones are waiting to be picked up. We tried our luck at Coyle Spring looking for green jasper and along Thronson Creek searching for vistaite. All we got at Coyle Spring was a flat tire where a sharp rock (not even green jasper) cut a small gash in our tread! A couple of dilapidated mines, abandoned long ago, are collapsed on the slopes southwest of Thronson Creek and form a nostalgic scene.

We recommend a long loop route west of U.S. Highway 26 for a good introduction to the rocks and geological formations in the Ochoco. Much of the route can be driven, with short hikes to the major attractions, or the entire loop can be hiked in several days. Take Forest Road 27 west from U.S. Highway 26 just below the Ochoco Divide Campground. There is an immediate climb through a dense conifer forest to the ridge. Hikers will want to take the first road left off of Forest Road 27 for a climb to the ridgetop. Continue on 27 to View Point for a spine-tingling overview of the Ochoco. From View Point the route continues through rich agate beds to Whistler Spring, a well-known site for Oregon's prized thunder eggs, rocks with unusual and colorful crystal formations inside. Take time to explore the rugged Desolation Canyon immediately west of Whistler Spring. It is another good thunder egg area. Continue west to Hash Rock for another superior view of the forest. A few miles south is the forest's remarkable Steins Pillar, an elongated rock pillar that stands 350 feet tall from a base 120 feet in diameter. It protrudes high above the surrounding trees and the two smaller columns nearby. East from Steins Pillar is White Fir Spring and White Rock, two further sources for thunder eggs. We blamed hundreds of previous rock hounds for our failure to find any thunder eggs here.

Another geological phenomenon is Twin Pillars, a pair of rocky columns reached by an eight-and-a-half-mile trail from the Wildcat Campground. Although the trail goes through scenic canyons and along bluffs, it is not a difficult one.

An old, undated Ochoco Forest map we were carrying mentioned excellent spring wildflower areas near Indian Prairie east of Delintment Lake. We found the areas teeming with wildflowers when we were there in mid-June. The three-petaled Mariposa lilies and sizzling orange Indian paintbrushes highlighted the display.

The Black Canyon Wilderness is another special forest attraction and is a small, wild, roadless area at the eastern edge about 25 miles northeast of Paulina. The 6,372-foot Wolf Mountain dominates this region of heavily timbered slopes and steep bluffs. Ponderosa pine and Douglas fir are the chief species of trees. A few miles northwest and reached by a gravel road is Spanish Peak. Since this is the highest point in the vicinity at 6,871 feet, there is a full-circle vista from the summit. We were told that there were about five dozen wild horses roaming in the Big Summit Ranger District north of Paulina, but we were not fortunate enough to see any during the short time we drove through the area.

Rogue River National Forest

Location: 584,292 acres in southwestern Oregon, north of Medford and Klamath Falls, a small part extending into northern California. State Routes 62, 140, and 230 cross the forest. Maps and brochures available at the district ranger stations in Jacksonville, Ashland, Butte Falls, and Prospect or by writing the Forest Supervisor, 333 W. 8th St., Medford, OR 97501.

Facilities and Services: Campgrounds (some with camper disposal service); picnic areas with grills; boat ramps; swimming beaches; winter sports facilities.

Special Attractions: Rogue River Gorge; Sky Lakes Area.

Activities: Camping, picnicking, hiking, backpacking, hunting, fishing, boating, swimming, winter sports.

Wildlife: Black bear, elk, bobcat.

THE WILD AND UNTAMED ROGUE RIVER has its origin in the northeast corner of the Rogue River National Forest, just a short distance north of the boundary of Crater Lake National Park. In a matter of a few miles, the river becomes a raging torrent and narrows with an ear-shattering roar through the Rogue River Gorge, which it has gouged out over the centuries. Park in the designated area alongside Route 62 and walk over to the gorge. At one point along the trail, all the trees, herbs, and rocks are covered with eerie-green mosses. The river unabashedly gushes through an opening in a vertical cliff. Three and a half miles south, reached either by hiking the Rogue River Recreation Trail or by driving Route 62, is Natural Bridge, where the Rogue River drops suddenly into a hole in volcanic rock, only to surge back to the surface a few feet below. There is a comfortable campground beneath tall sugar pines near the Natural Bridge.

Another four miles south of the Natural Bridge is the Mammoth Pines Nature Trail. The feature here is the fallen remains of a 500-year-old sugar pine that was 224 feet high and that had a diameter of 7 feet 11 inches. The trail circles a pretty woods of western hemlock, sugar pine, Douglas fir, grand fir, and ponderosa pine, but none of these living trees can match the girth of the fallen mammoth.

We particularly like the scenic drive along Route 230 to the Rogue River Gorge and then east on Route 62 to the west entrance of Crater Lake National Park. Pines and firs more than 125 feet tall tower above the road. A stop almost anywhere along the route puts you in still woods with moss-covered rocks. In mid-June the six-inch-wide white flowers of the Pacific dogwood brighten the forest. One of the best, although rugged, ways to see the forest trees is to hike from Abbott Creek to Abbott Butte. As you climb 2,800 feet in elevation from the creek to the top of the butte, you will see outstanding specimens of western hemlock near the creek, ponderosa pine, Douglas fir, sugar pine, western white pine, and incense cedar higher up, and finally Shasta red fir and mountain hemlock at the highest elevations.

From the southern boundary of Crater Lake National Park to Fourmile Lake is the Sky Lake Area, a region of 200 lakes shared with the Winema National Forest. Many of the lakes in the Rogue River part of the area are in the Seven Lakes Basin, a region where 7 sky-blue lakes are clustered together. The area is a must spot to visit in the forest. Several hiking trails lead to the area.

A favorite trail for hikers is to Mt. McLaughlin, the highest peak in the southern Cascades Range. Most hikers will stop short of the summit, since the peak should only be climbed by experienced mountain climbers. All but the last mile and a half of the trail are in the Winema National Forest.

The Applegate Valley a few miles north of the California border was gold country during the last two decades of the 19th century. Once the gold was panned out of the creeks, the miners used hydraulic mining to excavate gold from the hillsides. Remnants of this hydraulic mining can be seen along the fascinating Gin Lin Trail. Get a leaflet to give you more details, from a district ranger station, before starting the trail.

Siskiyou National Forest

Location: 1,060,178 acres in southwestern Oregon, extending into northern California, between Port Orford and Brookings on the west and Grants Pass on the east. U.S. Highways 101 and 199 are the access roads. Maps and brochures available at the district ranger stations in Brookings, Grants Pass, Gold Beach, Cave Junction, and Powers or by writing the Forest Supervisor, Box 440, Grants Pass, OR 97526.

Facilities and Services: Campgrounds (some with camper disposal service); picnic areas with grills; boat ramps; swimming beach.

Special Attractions: Kalmiopsis Wilderness; special botanical areas.

Activities: Camping, picnicking, hiking, backpacking, hunting, fishing, boating, swimming, horseback riding, nature study.

Wildlife: Bald eagle, mountain lion, bobcat, black bear, osprey, Columbian black-tailed deer.

AFTER NOTICING on the Siskiyou map a statement that the forest was a "botanist's paradise" with places like Babyfoot Lake Botanical Area and York Botanical Area, this writer, who is a professional botanist, was eager for his first visit to the forest. To top it all, there was the 169,000-acre Kalmiopsis Wilderness, named especially for a rare pink-flowered shrub of the heath family that grows there. With all these promised goodies in store, we headed for the wilderness by taking a most difficult, twisting, narrow dirt road to Chetco Pass at the northeastern corner of the area. There are trails that lead north, south, and west from Chetco Pass, and all should be taken if time and stamina permit. The trails are rugged and pass through areas of poison oak. Rattlesnakes are said to be common, particularly around Eagle Mountain. We headed west on the Upper Chetco River Trail, which drops continuously for nearly 2,000 feet into the wilderness, first past a couple of abandoned mines before coming, after three miles, to a nice stand of the rare Kalmiopsis near Slide Creek. The shrubs, with their rosy pink flowers, look like pint-sized rhododendrons and, except for a small group of them along the Umpqua River to the north, are confined to the area in and around the Kalmiopsis Wilderness.

We backtracked to Chetco Pass, then hiked north toward Pearsoll Peak through the reddest terrain imaginable. The red is peridotite, a strange type of volcanic rock among which grow dwarfed trees and rare shrubs and wildflowers. As you hike across the red rocks to Pearsoll Peak, about the only vegetation you will see are low-growing junipers and manzanitas. For the next two miles the trail passes through a forest of Jeffrey pines to Pearsoll Peak where there is a full-circle view of the Siskiyou Range. We wanted to continue on another two miles past the nearly bald Gold Basin Butte to Granite Springs, because another rare member of the heath family, called the Siskiyou leucothoe, grows there. Although this plant's little white, bell-shaped flowers had already bloomed in May, we enjoyed seeing it as well as the thick stand of conifers nearby, which included the delightful sugar pine with its 15-inch-long cones.

Returning to Chetco Pass, we took the seven-mile trail south to Babyfoot Lake by way of Eagle Mountain, Eagle Gap, Whetstone Butte, and Onion Camp. From the bald peak of Eagle Mountain, the trail drops down to Eagle Gap, where there are fine specimens of the large-coned Jeffrey pine. Beyond Whetstone Butte and more vivid red peridotite rocks is Onion Camp, named for masses of a showy wild onion that grows all over the place. Two miles on is Babyfoot Lake, set in the midst of the Babyfoot Lake Botanical area. The shallow, deep blue lake is nestled in front of a sheer rock cliff. The botanical delight around the lake is one of the densest stands of Brewer's weeping spruce, one of the nation's rarest trees. You will recognize it readily because of its hanging branches. If you wish to hike for several more days, follow the trail along the Chetco River which borders the southern edge of the wilderness, coming at last to Vulcan Peak where you might want to arrange to be picked up at the end of Forest Road 3917.

The York Creek Botanical Area can be reached by a rough 2-mile trail along the wild Illinois River from the bridge over Briggs Creek about 15 miles northwest of Selma. When you get to the botanical area, you will find a small stand of Kalmiopsis, a rare tree known as the Port Orford cedar, and the one-of-a-kind cobra plant, an insect-eating type of pitcher plant that is shaped like a cobra ready to strike, complete with hood and forked "tongue."

There is a better stand of Port Orford cedars between Johnson Creek and the South Fork Coquille River. It is 18 miles south of Powers, just south of the junction of Forest Roads 326 and 333. Some of the oldest Port Orford cedars in the grove are at least 175 feet tall and more than 8 feet in diameter, but they are exceeded in height by 200-foot-tall Douglas firs.

Another remote but spectacular area featuring several spectacular plants occupies a steep north-facing slope adjacent to Wheeler Creek, about 18 miles east of Brookings via a series of paved and graveled roads. The botanical prize here is one of the northernmost stands of giant redwoods, some of them topping 200 feet and having a trunk diameter of 8½ feet. The area is extremely rugged with dense vegetation, and there are no developed trails. For an easier look at Oregon redwoods, take the Redwood Nature Trail just north of the Alfred A. Loeb State Park. Although the trail is steep in places, there is a well-beaten path that passes through this lush, near-subtropical forest. In addition to many redwoods—one of them is 286 feet tall and 12 feet 6 inches in diameter—there are giant Douglas firs, purple rhododendrons that bloom in mid-June, big-leaf maples, Oregon myrtles, and tan oaks. An abundance of sword fern, lady fern, deer fern, and maidenhair fern fills in the understory.

The Rogue River cuts across the forest from Marial to Gold Beach; at some places it is so rough and undisturbed that a part of it has been designated a national Wild and Scenic River. One popular way to see some of the Rogue is to take commercial jet boat rides from Gold Beach to Agness or Paradise Bar. The 64-mile round trip to Agness is on peaceful waters crowded with fishermen; ospreys and Columbian black-tailed deer are usually seen on every trip. The 104-mile round trip to Paradise Bar gets into whitewater rapids, but the trip is reasonably safe. The stretch of the Rogue River between Paradise Bar and Marial, however, is only for the experienced rafter since it goes through dangerous Mule Creek Canyon, a 20-foot-wide gorge enclosed by 50-foot-tall vertical cliffs. The Rogue River Hiking Trail parallels the river throughout most of its length.

Siuslaw National Forest

Location: 628,442 acres in western Oregon, along the coast from North Bend to Cape Lookout. U.S. Highway 101 runs the length of the forest and State Routes 18, 34, 36, and 126 cross inland. Maps and brochures available at the district ranger stations in Alsea, Hebo, Mapleton, and

Waldport, at the Oregon Dunes National Recreation Area headquarters at Reedsport, at the Cape Perpetua Visitor's Center, or by writing the Forest Supervisor, Box 1148, Corvallis, OR 97339.

Facilities and Services: Campgrounds (some with camper disposal service); picnic areas with grills; boat ramps; swimming beaches; visitor's centers.

Special Attractions: Oregon Dunes National Recreation Area; Cape Perpetua.

Activities: Camping, picnicking, hiking, beachcombing, dune buggying, boating, swimming, hunting, fishing, nature study, whale watching.

Wildlife: Black bear, bobcat.

THE 50 MILES of ocean shoreline, capped by the formidable Umpqua Sand Dunes, make the Siuslaw of western Oregon one of the most unusual of the 153 national forests in the United States. The Umpqua Dunes are the highest of the many dunes that comprise the Oregon Dunes National Recreation Area, a star attraction of the forest. The sand of this long stretch of dunes, which sometimes extends 3 miles inland, originally came from Oregon mountain sandstone that washed out to sea, only to be blown back and piled up by strong westerly winds.

You may want to make your first stop at the visitor's center in Reedsport to learn about the dunes and to plan your activities on the sand. We made our first foray from the Eel Creek Campground near Lakeside, where a nature trail leads to the edge of the Umpqua Dunes. The going is tough if you try to climb to the top of the highest dunes, since it is easy to slide back two steps for every single step forward. From the top of the 200-foot dunes you can see the waves of the Pacific Ocean breaking against the shoreline a mile away. Here and there are small, isolated pools of water that have collected in shallow sandy basins. An occasional beach grass or a purple-flowered lupine has somehow established a foothold in the otherwise continuous sea of sand.

Hiking is a slow and often tedious way to travel over the dunes, but it is the only way in the areas where vehicles are not permitted. Dune buggies are allowed in some areas, however, and can offer several thrills a minute. You can ride with a commercial operator, or bring your own dune buggy. If you drive your own, make sure you understand the many hazards that await you in the sand.

It is fun and relaxing to walk along the beach, letting the tide sweep across your feet with its pulsating rhythm. Among the debris you might find are colorful agates, seashells, old bottles, and driftwood, each piece shaped differently. On our last trip we found an old Johnson & Johnson talcum can with a Japanese label, which may have made the long trip to Oregon from the Orient. But it is the sandblasted driftwood that caught our attention, for these pieces of dead, gnarled, bleached pines have been polished smooth by the scouring action of the sand.

If you tire of the sand, fish for a while in the freshwater lakes that dot the landscape inland from the dunes. Alder Lake is one and offers a comfortable nearby campground. The lake is near the upper end of the strip of dunes.

North beyond Alder Lake, Highway 101 follows an often precipitous route along the ocean, eventually coming to the wild and often turbulent Cape Perpetua area, another major forest attraction. You should allow a few hours to explore this region. Begin at the modern visitor's center and watch a short film that describes the wonders of the cape, which is one of the highlighted regions of the forest. Then walk the trails to the ocean to observe for yourself all the various currents and tides, including the spouting horns, tide pools, the violent Devil's Churn, and the tranquil Cape Cove Beach. We were entranced watching the streams of water of the spouting horns being forced upward through small openings in the rocks. At another place on a rocky shelf, foaming ocean water erupts from a large, circular chasm, then quickly drains back into the cavity, only to reappear a few seconds later to start the process over again. From these turbulent water displays, the trail abruptly enters Cape Cove Beach, a sandy cove protected from the action of waves. This is a good place to seek driftwood and the delightful and tasty razor clams. During low tide, small pools of water, aptly termed tide pools, become exposed and offer a living marine laboratory. You will probably see a spiny purple sea urchin, or an intricately shaped sea anemone, or a hermit crab, or a starfish among a myriad of sea life. Observe, but do not molest, because it is illegal to take specimens from this delicate habitat.

Near the northern end of the forest is a conspicuous promontory that juts out above the ocean, known as Cascade Head. Take time to explore this region by absorbing the mighty views and walking in the rugged forest on the crest. It is one of the best places along the coast to see Sitka spruce and western hemlock. A few of the spruces are nearly 250 feet tall and have a trunk diameter of 7 feet. If you decide to explore here, you must exercise caution because of the many steep cliffs that drop more than 1,000 feet to the ocean.

Umatilla National Forest

Location: 1,088,158 acres in northeastern Oregon and a small portion of southeastern Washington, south and east of Pendleton, Oregon, and southeast of Walla Walla, Washington. Oregon Routes 82 and 204 are the major roads in the forest. Maps and brochures available at the district ranger stations in Dale, Heppner, and Ukiah, Oregon, and Pomeroy and Walla Walla, Washington, or by writing the Forest Supervisor, 2517 SW Hailey Ave., Pendleton, OR 97801.

Facilities and Services: Campgrounds (some with camper disposal service); picnic areas with grills; boat ramps; swimming beaches.

Special Attractions: Wenaha-Tucannon Wilderness.

Activities: Camping, picnicking, hiking, backpacking, hunting, fishing, horseback riding, boating, swimming, winter sports, nature study, rafting.

Wildlife: Black bear, elk, bobcat.

IN A REMOTE PART of Oregon and Washington, where the Blue Mountains span the border between these states, is the Wenaha-Tucannon Wilderness, 176,800 acres of rugged ridges of basalt, vertical cliffs, and fast-flowing streams. The Wenaha River has formed a deep canyon rimmed by dense forests. A trail that follows the entire length of the river gives a good cross section through the wilderness.

South of the Godman Guard Station is a two-mile trail to Rainbow Creek. When you get to the creek, climb the 1,100 feet from the creek to the summit of Sugarloaf Butte. You will see some of the best stands of uncut grand fir, Douglas fir, western white pine, ponderosa pine, and western larch in the state of Washington. It is not unusual to see an elk grazing near the creek. Another recommended trail in the wilderness is the three-miler from the Teepee Campground east of the Godman Guard Station to Oregon Butte, the highest elevation in the wilderness at 6,401 feet. From the summit of Oregon Butte is a full-circle panorama of the wilderness.

There are several vista points where you can peer into the wilderness without hiking the trails. Sunset Point, about 30 miles south of Pomeroy, Washington, offers a sterling view south of the Tucannon River. Table Rock, a high rock along the spectacular but primitive Skyline Drive, offers an outstanding view into the western side. Our favorite is the one from Big Rocks that overlooks the Wenaha River Canyon.

The southeast edge of the Umatilla is in gold country, and several closed mines, abandoned towns, and rusted equipment can be seen here and there. Eight miles west of Granite, Oregon, is the Fremont Powerhouse, a registered Historical Place. The powerhouse was constructed in 1908 to provide electrical power for nearby gold mines. It stayed in use until 1968. All of the buildings, pipelines, and even the instrument panel are in place. Another nine miles west of the powerhouse is glacier-carved Olive Lake, the only natural lake in the forest. It is reported to be excellent for fishing. There is a campground nearby.

For a white-water experience, raft the 35 miles of the Grande Ronde River from Rondowa to Troy. You are sure to get a jolt or two.

Umpqua National Forest

Location: 989,144 acres in southwestern Oregon, east of Roseburg and Canyonville. State Route 138 crosses the forest. Maps and brochures available at the district ranger stations in Cottage Grove, Tiller, Idleyld Park, and Glide, at the Diamond Lake Information Center, or by writing the Forest Supervisor, Box 1008, Roseburg, OR 97470.

Facilities and Services: Campgrounds (some with camper disposal service); picnic areas with grills; boat ramps; swimming beaches; winter sports facilities.

Special Attraction: Diamond Lake Recreation Area.

Activities: Camping, picnicking, hiking, backpacking, hunting, fishing, boating, swimming, winter sports.

Wildlife: Black bear, mountain lion, bobcat.

DIAMOND LAKE is sandwiched between Mt. Thielsen and Mt. Bailey a few miles north of Crater Lake National Park. You should stop at the visitor's information center on the east side of the sparkling lake to find out all the recreational opportunities in the area, for this lake is the forest's prime visitor attraction. There are several boat launching ramps, a couple of comfortable campgrounds, a camp store, and a short nature trail. One morning when we were there in a heavy mist, a flock of more than 20 bright yellow western tanagers flitted about in a small tree for half an hour.

From Diamond Lake you can see the prominent spirelike peak of Mt. Thielsen. Early settlers called the mountain Big Cowhorn, a descriptive name well chosen for the peak. A five-mile trail from Diamond Lake to Mt. Thielsen climbs 3,982 feet. It is steep in several places, but only the last segment is difficult.

The highway that parallels swift-flowing North Umpqua River from Roseburg to Diamond Lake provides a scenic cross section of the forest. Near Glide, the North Umpqua and Little rivers join at right angles, an unusual phenomenon which is noted as the Colliding Rivers on the signpost above this junction. We were amazed by a geological formation along the North Umpqua River known as the Basaltic Columns. Crowded perpendicular pillars of volcanic rock known as basalt rise several feet from the edge of the river. There are several waterfalls in the forest adjacent to the North Umpqua Highway that are worth a visit. Most spectacular is Watson Falls which drops 272 feet into a catch basin below. It is reached by a half-mile trail from the Watson Falls Picnic Area. Other falls are Toketee, Clearwater, Whitehorse, and Fall River.

Plant lovers are sure to enjoy a 40-acre stand of some of the biggest incense cedars in the United States. These strong-scented trees about 11 miles west of Diamond Lake are reached by a series of confusing back roads. It is best to seek directions from a district ranger.

Wallowa-Whitman National Forest

Location: 2,251,031 acres in northeastern Oregon, on either side of Baker and La Grande. Interstate 84 and State Routes 7, 82, and 86 are the major roads. Maps and brochures available at the district ranger stations in Baker, Enterprise, La Grande, Halfway, Union, Unity, and Joseph or by writing the Forest Supervisor, Box 907, Baker, OR 97814.

Facilities and Services: Campgrounds (some with camper disposal service); picnic areas with grills; boat ramps; swimming beaches; winter sports facilities.

Special Attractions: Hells Canyon; Eagle Cap Wilderness.

Activities: Camping, picnicking, hiking, backpacking, hunting, fishing, horseback riding, boating, swimming, winter sports, nature study.

Wildlife: Black bear, elk, mountain goat, rare Wallowa gray-crowned rosy finch.

FOR AEONS the Snake River has carved the deepest gorge in the United States. Hells Canyon, which plunges to a depth of 7,000 feet, separates Oregon from Idaho and the Wallowa-Whitman National Forest from the Nezperce and Payette forests. A good place to start exploration of the Wallowa-Whitman side of Hells Canyon is in the canyon itself, a star forest attraction. A primitive road north from Homestead parallels the impounded Snake River for a while until it runs out of room and narrows into a hiking trail. Skilled hikers can follow the river all the way to Hells Canyon Dam. During the summer the temperature in the canyon is sweltering, so that when the trail crosses one of the scenic tributaries into the Snake River, such as Steamboat Creek, it is a welcome respite from the arid conditions.

We also recommend seeing Hells Canyon from above, and the best place to do it is from Hat Point, an isolated perch 7,000 feet above the Snake River. To reach Hat Point from the canyon requires an arduous all-day drive through the Wallowa-Whitman, some of it by way of a hairy-scary dirt and gravel road. The vista from Hat Point makes the ordeal worthwhile. The road to Hat Point from the canyon leaves Route 86, 7 miles west of Copperfield. All the way to Imnaha, the road follows the course of first one creek or river and then another. If you wish to linger along the way, there are campgrounds at North Pine Creek and the Imnaha River. After reaching the community of Imnaha, it is still 24 miles to Hat Point by way of a one-lane road that is usually open by July 1. Careful drivers will have no problem with the road as it snakes its way over dry ridges and around vertical walls. There are several places to pull out and view the surroundings, but the best panorama awaits you from the lookout tower at Hat Point. You can also stretch your legs by hiking the trails that radiate from the lookout.

In sharp contrast to Hat Point and 20 miles west is another forest special attraction, Eagle Cap Wilderness, a 221,000-acre wild area of snowy mountain peaks, densely forested ravines, glacial lakes, and bubbling streams. Broad-topped Eagle Cap Peak, which has lent its name to the wilderness, towers above a string of gemlike lakes. Trail #1810 skirts Horseshoe, Lee, Craig, Crescent, Douglas, Sunshine, and Mirror lakes before passing along the western slope of Eagle Cap. Another feature attraction is the Matterhorn, a conspicuous white limestone mountain that reaches 9,595 feet. As a treat for bird-watchers, the rare Wallowa gray-crowned rosy finch, a small bird with contrasting colors, nests only in this wilderness.

One of the more rewarding outings in the forest is the 106-mile, round-trip Elkhorn Drive from Baker through country steeped in history and rich in scenery. The trip can be made easily in a full day, but you may wish to pause awhile and extend your trip a day or two. There are several convenient campgrounds along the way. Pick up an Elkhorn Drive road guide at one of the ranger stations in the forest before you start. Much of the first part of the road from Baker to Sumpter follows the route of an old narrow-gauge railroad that hauled pine logs from forest to mill. Gold, discovered along the Powder River in 1862, led to a great gold rush, and evidence of past mining operations can still be seen in the form of abandoned equipment, piles of boulders, gravel, and black slag, and remains of towns, such as Bourne and Gold Center. About a mile north of Granite we were fascinated by rock walls in the stream bottom next to the road. These were hand built by Chinese miners who, more than a century ago, piled up the larger boulders so that they could seek gold in the finer gravel and sand that lay beneath. After crossing the North Fork of the John Day River, the road turns abruptly east toward Baker, but the most dramatic scenery is yet to come. High in the Blue Mountains is beautiful Anthony Lake, resting at the foot of 8,366-foot Gunsight Mountain, the mountain with a notched peak. When the sun is right, the mountain is mirrored in the water. When we were there in mid-June, the lake was still snow covered. Earlier in the year, the area is buzzing with winter sports enthusiasts. When the snow melts by early July, the educational Trail of the Alpine Glacier is exposed. This short trail leads from the south end of Anthony Lake and passes granite boulders brought in by glaciers, glacial lakes that have developed in gouged-out depressions, hanging valleys which are glacier-carved side canyons that have been left hanging above the main canyon, and a U-shaped glacial valley. A guide leaflet is available to explain these phenomena.

Willamette National Forest

Location: 1,675,000 acres in central Oregon, east of Albany and Eugene. U.S. Highway 20 and State Routes 22 and 126 cross the forest. Maps and brochures available at the district ranger stations in Blue River, Detroit, Lowell, McKenzie Bridge, Westfir, Oakridge, and Sweet Home or by writing the Forest Supervisor, Box 10607, Eugene, OR 97440.

Facilities and Services: Campgrounds (some with camper disposal service); picnic areas with grills; boat ramps; swimming beaches; winter sports facilities.

Special Attractions: Mt. Jefferson Wilderness; Mt. Washington Wilderness; Three Sisters Wilderness; Diamond Peak Wilderness.

Activities: Camping, picnicking, hiking, backpacking, hunting, fishing, horseback riding, boating, swimming, nature study, winter sports.

Wildlife: Black bear, elk, wolverine.

THE WILLAMETTE is a land of endless stretches of trees. Forest service personnel note that about 10 percent of all timber cut in the national forests comes from the Willamette. Despite the heavy emphasis on the commercial use of the forest, there is still plenty of opportunity for recreation.

For starters, there are the four wilderness areas—Mt. Jefferson, Mt. Washington, Three Sisters, and Diamond Peak—each named for the prominent mountains within them. The wildernesses are shared by the Deschutes Forest, and Mt. Jefferson is also partly in the Mt. Hood Forest. Mt. Jefferson, with its huge white peak, ranks among the most beautiful mountains in the world. There are many deep blue lakes in the wilderness, for which Mt. Jefferson serves as a perfect backdrop. Mowich Lake is unusual in that it has a heavily forested island within it. Marion Lake is popular because it is only two miles from the wilderness boundary and is quickly accessible. From Marion Lake is an unsurpassed view of Three Fingered Jack Mountain. Our favorite trail is the one to Pamelia Lake, which passes through a dense forest of skyscraper Douglas firs. North and South Cinder peaks are two smaller volcanic cones in the wilderness.

Although Mt. Washington is the dominant feature of the Mt. Washington Wilderness, vast plains of lava that cover one-third of the wilderness make it a desolate landscape. Belknap Crater is in the heart of the lava plains. Most of the trails, as you might expect, are exceptionally rough.

Three Sisters Wilderness is wild, wild mountainous country with the imposing Sisters the star attractions, each with active glaciers mounted on their upper shoulders. In the southern end of the wilderness are dozens of tranquil blue lakes, many of them awaiting the angler's line. Almost every lake is surrounded by dense, green, narrow-spired conifers.

Diamond Peak Wilderness is southernmost in the Willamette. The Pacific Crest National Scenic Trail follows the complete crest of the Cascades through all of the wildernesses in the forest. There is a good 10-mile trail along the western edge from Hemlock Butte to Ruth Lake, which offers several views of Diamond Peak.

You do not have to be a backpacker or long-distance hiker to enjoy the Willamette. The McKenzie Highway, Route 126, is a breathtaking road through the forest, but since it goes over a heavy snow belt at McKenzie Pass, it is only open a few months of the year. Stop awhile at the pass, which is in the middle of a lava field, and have a good look. The National Aeronautics and Space Administration (NASA) uses the area to determine the ability of astronauts to travel on lava. Surprisingly, dwarf "bonsai" trees of lodgepole and ponderosa pine, mountain hemlock, and alpine, silver, and noble firs grow among the lava rocks.

Another area where you will need to reserve some time is around Clear Lake, a body of water so clear that you can see to a depth of 100 feet or more. At one end of the lake is a submerged forest that was inundated more than 1,000 years ago during a lava flow. South of Clear Lake are three rushing waterfalls worth a visit—Sahalie, Koosah, and Tamolitch.

Waldo Lake, with 6,700 acres of clear, blue water, lies high on the slopes of the Cascades. While there are modern campgrounds and boat ramps along the eastern side of the lake, the west and north shores are purposely undeveloped and primitive so that backcountry experience can be gained here.

Winema National Forest

Location: 1,043,179 acres in south-central Oregon, north of Klamath Falls and east of Crater Lake National Park. U.S. Highway 97 and State Routes 138 and 140 are the major roads. Maps and brochures available at the district ranger stations in Chemult, Chiloquin, and Klamath Falls or by writing the Forest Supervisor, Box 1390, Klamath Falls, OR 97601.

Facilities and Services: Campgrounds (some with camper disposal service); picnic areas with grills; boat ramps, winter sports facilities.

Special Attractions: Sky Lakes Area; Mountain Lakes Wilderness.

Activities: Camping, picnicking, hiking, backpacking, hunting, fishing, horseback riding, boating, winter sports, nature study.

Wildlife: Bald eagle, black bear, bobcat, mountain lion.

IN AN AREA only 6 miles wide and stretching for 27 miles from the southern boundary of Crater Lake National Park to Fourmile Lake are 200 lakes that make up the fabulous Sky Lakes Area. The western half of the region is in the Rogue River Forest, while the eastern half is in the Winema. There are plenty of lakes in both forests. The best trail to bring you near many of the lakes is the famous Pacific Crest Trail which follows the ridge of the southern Cascades throughout the entire length of the Sky Lakes. Several of the larger lakes have been stocked with trout and are popular with anglers. All of the lakes are scenic, and photographers have no trouble shooting up a lot of film.

At the north end of Sky Lakes, just beneath Crater Lake's southern border, are a pair of interesting volcanic cones known as Goose Nest and Goose Egg, which they amusingly resemble. Their existence is related to the activity of Mt. Mazama, the volcano responsible for Crater Lake.

Two large and popular lakes for water sports enthusiasts are outside of the Sky Lakes Area. Fourmile Lake and Lake of the Woods, both natural lakes at elevations of 5,700 and 4,950 feet, respectively, are ideal for camping, picnicking, boating, and swimming.

South of the Sky Lakes is another Winema attraction— the Mountain Lakes Wilderness, a perfectly square, 36-square-mile primitive area with its own group of blue mountain lakes. The lakes are nestled among Whiteface Peak, Greylock Mountain, Mt. Harriman, Mt. Carmine, and Crater Mountain, all peaks in the 7,500-to-8,000-foot range.

There are several geological features in the forest that are worth a visit. Several fields scattered with pumice, a lightweight volcanic rock, can be found, mostly in the eastern part. At one area west of Bluejay Spring, the pumice, which is 1 to 10 feet deep, supports a plant community dominated by ponderosa pine, the shrubby bitterbrush, and needle grass; the prominent wildflowers are a purple-flowered violet and one of the false buckwheats.

A hodgepodge of amusing rock formations is found at the Devil's Garden northeast of Klamath Falls. You can supply your own names to the diverse formations. East of Crater Lake National Park is a group of pointed rocks along Sand Creek known as the Sand Creek Pinnacles. One curious phenomenon is known as Mares Egg Spring. The "eggs" are huge spherical globs of *Nostoc*, a microscopic aquatic plant that has massed together in large round colonies. They appear as green stones in the bottom of the spring.

Colville National Forest

Location: 944,779 acres in northeastern Washington, on either side of Colville and the Columbia River, and extending to the Canadian border. U.S. Highway 395 and State Routes 20, 21, and 31 serve the forest. Maps and brochures available at the district ranger stations in Colville, Kettle Falls, Newport, Republic, and Metaline Falls or by writing the Forest Supervisor, 695 S. Main St., Colville, WA 99114.

Facilities and Services: Campgrounds (some with camper disposal service); picnic areas with grills; boat ramps; swimming beaches.

Special Attractions: Sullivan Lake; Chewelah Mountain.

Activities: Camping, picnicking, hiking, hunting, fishing, boating, swimming, winter sports.

Wildlife: Bald eagle, black bear, mountain goat, mountain lion.

THERE IS a strong sense of serenity when you visit the Colville. The roads and trails pass through winding verdant valleys and over soothing and rolling wooded slopes, past small blue lakes and clear white streams.

Sullivan Lake is a major recreation area and forest attraction on the eastern side of the forest near the community of Metaline Falls. There are several choice campsites along the lake where you can fry your day's catch before falling asleep beneath whistling pines. There also are several hiking trails you can take, including one that climbs to Sullivan Mountain a few miles north of the lake.

The most spectacular feature in the vicinity of Metaline Falls is Peewee Falls, a 200-foot ribbon of white water that cascades down a vertical cliff and splashes into Boundary Dam Reservoir. The falls is best seen from a boat in the reservoir. If you try to get a view of the falls from adjacent hilltops, be careful, for the footing is treacherous. There is also an attractive 35-foot falls along Marble Creek about 15 miles northeast of Colville.

East of the small community of Chewelah is Chewelah Mountain, a popular forest attraction during all seasons. We enjoyed hiking there in early summer. Later in the year winter sports enthusiasts flock to the area.

Half of the Colville lies west of the Columbia River. This mountainous region, whose tallest peak only reaches 7,135 feet, seems mild compared with mountains of the northwest, despite such formidable names as Profanity Peak, Jungle Hill, and Storm King Mountain. The route along Sherman Creek, which climbs over Sherman Creek Pass between Snow Peak and Jungle Hill, is a good one either to hike or drive. The scenery is among the finest the forest has to offer.

Gifford Pinchot National Forest

Location: 1,252,895 acres in south-central Washington, between Yakima and Vancouver, Washington. U.S. Highway 12 and State Route 14 are the major access roads. Maps and brochures available at the district ranger station in Amboy, Trout Lake, Packwood, Randle, and Carson, at the Mt. St. Helens Visitor's Center in Lewis and Clark State Park, or by writing the Forest Supervisor, 500 W. 12th St., Vancouver, WA 98660.

Facilities and Services: Campgrounds (some with camper

disposal service); picnic areas with grills; boat launch areas; swimming beaches; visitor's center.

Special Attractions: Mt. St. Helens Area; Mt. Adams Wilderness; Goat Rocks Wilderness.

Activities: Camping, picnicking, hiking, backpacking, hunting, fishing, boating, swimming, horseback riding, berry picking, winter sports facilities.

Wildlife: Mountain goat, black bear, elk, mountain lion.

A PART OF the Gifford Pinchot National Forest changed drastically on May 18, 1980, when, at 8:32 A.M., the volcano Mt. St. Helens erupted, causing widespread devastation. Mounted high on the western edge of the forest, St. Helens destroyed vast areas of valuable timber, wiped out lakes and streams, and took the lives of several humans and countless wildlife. Visitors to the forest should make a point of stopping at the visitor's center the forest service has erected in Lewis and Clark State Park. There you can watch films and slide shows of the eruption and find out the best place to view the resulting devastation.

The area immediately around Mt. St. Helens is still off limits to visitors, but you can get close enough to see what havoc was wrought. Drive south from Randle on Route 25. After several miles turn west on Route 99 and make a stop at Bear Meadow where you can see Mt. St. Helens, which is about 11 miles away. It was here that photographer Gary Rosenquist took his now-famous sequence of pictures of the volcano erupting. The area is just outside the blast-damage area. Continue on Route 99. As we rounded one curve, we suddenly got our first glimpse of the blast area. It was an undoing experience. Before us lay nothing but gray desolation. A forest of trees, now all ashen in color, lay on its side. No living creatures moved; no birds sang. More striking was the fact that where our road abruptly entered the blast zone, the area immediately behind us was still green and alive with sights and sounds of nature.

Route 99 continues on for a few miles. It passes a miner's car that was damaged by the blast. Farther on, we parked our car and took the 200-yard walk to Meta Lake, where dangerous bacteria still lurk so that the water should not even be touched, let alone drunk. Another mile ahead, where

Route 99 junctions with 92, there is another place to view Mt. St. Helens, only seven and a half miles distant.

We backtracked to Route 25, then headed south toward Cougar. We stopped at Muddy River, whose total character was changed when the volcano erupted. Huge boulders were washed into the river channel from many miles away; streamside vegetation was flattened and killed.

Two miles farther south, at the Cedar Flats Research Natural Area, huge western cedars tower above giant ferns, showing no evidence of St. Helens' violence. A short trail winds through this tropiclike forest, but there is evidence of volcanic eruptions centuries ago in other parts of the forest. Ape Cave, the longest lava tube in the United States, is southwest of Mt. St. Helens. You will need a lantern if you plan to walk into the tube through which lava flowed long ago. Nearby is a lava cast forest, where only the shells are left of trees that were encased in lava. At the southeastern corner of the Gifford Pinchot, in an area 10 to 20 miles north of the Columbia River, is the 12,500-acre Big Lava Beds, where piles of lava, short caves, and a variety of strangely shaped lava forms prevail. The volcanic cone responsible for this massive flow lies in the northern portion of the area. Forest service roads completely encircle the flow.

Mt. Adams, with its glorious white 12,326-foot peak, dominates the eastern edge of the forest. It is also the central feature in the 42,411-acre Mt. Adams Wilderness, a major drawing card of the forest. A good base from which to explore the wilderness is the Bear Creek Meadows Campground just outside the southern boundary of the wilderness. Bear Creek Meadows are, themselves, a spectacular sight of colorful mountain wildflowers. The short Trail of the Flowers winds through this collection of native alpine plants; signs along the way point out the species.

Located midway between Mt. Rainier and Mt. Adams is a region of alpine meadows, rocky spires and pinnacles, swift streams, and glaciers known as Goat Rocks Wilderness. It, too, is a forest special attraction. Much of it is above timberline, and heavy snow covers many of the trails until mid-July. Be sure and carry a wildflower guide with you to identify the myriad of alpine wildflowers that bloom during a two-month period. One of the most rewarding hikes is over the Lily Basin Trail. From it are stunning views of Packwood Lake with Mt. Rainier behind. A part of the Goat Rocks Wilderness is administered by the Wenatchee National Forest.

Indian Heaven is a secluded area of low peaks and dozens of mountain lakes that is popular among hikers and fishermen. At the south end of Indian Heaven is the historical Indian Racetrack where pony races were held centuries ago. A pony-worn groove 10 feet wide and 2,000 feet long is still visible. North of Indian Heaven are the Sawtooth Huckleberry Fields where huckleberry picking is at its finest.

Mt. Baker–Snoqualmie National Forest

Location: 2,506,988 acres in northwestern Washington, north and east of Seattle. Interstate 90, U.S. Highways 2 and 410, and State Routes 20 and 164 are the major roads. Maps and brochures available at the district ranger stations in Concrete, Darrington, Glacier, Granite Falls, North Bend, Skykomish, and Enumclaw, Washington or by writing the Forest Supervisor, 1022 First Ave., Seattle, WA 98104.

Facilities and Services: Campgrounds (some with camper disposal service); picnic areas with grills; boat ramps; swimming beach; winter sports facilities.

Special Attractions: Glacier Peak Wilderness; Alpine Lakes Wilderness.

Activities: Camping, picnicking, hiking, backpacking, hunting, fishing, horseback riding, boating, swimming, winter sports.

Wildlife: Bald eagle, black bear, elk, mountain lion, bobcat, mountain goat.

FROM THE Canadian border to the northern boundary of Mt. Rainier National Park and lying west of the crest of the Cascades Range, the Mt. Baker–Snoqualmie National Forest sprawls over nearly two million acres of high mountains, alpine meadows, glacial lakes, plunging waterfalls, and rushing streams.

Mt. Baker, at 10,778 feet, dominates the northern end

of the forest and keeps volcano watchers busy by its occasional belches of steam and smoke. There are splendid views of the mountain from Heather Meadow. There are also several kinds of wildflowers in the meadow that are indigenous only to this area. Take a wildflower guide along so you can identify these delicate beauties.

Glacier Peak Wilderness is a star attraction in the forest. It is 464,219 acres in size and is noted for its glacier-packed mountains. There are more active glaciers here than in any other area of the United States except Alaska. The wilderness, which extends into the Wenatchee Forest, has many other outstanding features including densely forested valleys of tall trees, a myriad of deep blue lakes, noisy waterfalls, roaring rivers, and meadows teeming with wildflowers. For one of the best wildflower meadows in the forest and easily accessible in a day, take the three-mile trail from the White Chuck trail head to Meadow Mountain. In July and August the meadow is completely carpeted with red, blue, yellow, and white blossoms.

Another wilderness shared with the Wenatchee is Alpine Lakes Wilderness, a prime visitor attraction. There are many sharp pinnacles and jagged spires projecting throughout the 306,000-acre wilderness. Many of the lakes are conveniently within a few miles of the trail heads. One such is Copper Lake—use the trail from Foss River Campground.

We also enjoyed the forest's ice caves found on the mile-long Ice Caves National Recreation Trail. The trail goes through a dense forest of western hemlock, western red cedar, and silver fir to the north face of Big Four Mountain where the fields are located. The ice caves are at the base of the mountain and are actually large tunnels caused by meltwater and warm air. The caves, however, are extremely dangerous and should not be entered.

Another trail worth taking is to Lake Twentytwo, a 44-acre glacial lake that is more than 50 feet deep. There is a permanent snowfield in the lake basin. The woods surrounding the lake are composed of fine old specimens of western red cedar and western hemlock. The trail to the lake, which leaves State Route 22 just east of the Verlot Ranger Station, passes a bubbling waterfall on Twentytwo Creek.

The first wagon road over the Cascades to the developing

villages on Puget Sound came across Snoqualmie Pass in 1868. Although Interstate 90 uses the same pass today to cross the mountain, about one mile of the old wagon road has been preserved. Pick up a Snoqualmie Pass Wagon Road booklet from a ranger station and take the trail. You will see several remnants of the old road, including some split cedar planks, called puncheons, used as a land bridge.

In the southern end of the forest, State Route 164, before reaching Mt. Rainier National Park, passes through a forest of giant Douglas firs. The Dalles Campground is nestled among the tall trees, and one mammoth tree is at the north end of the campground. Adjacent to the campsites is a nature trail through a rich woods known as the John Muir Grove.

Okanogan National Forest

Location: 1,499,498 acres in north-central Washington, northwest of Okanogan. U.S. Highway 97 and State Route 20 are the major roads. Maps and brochures available at the district ranger stations in Tonasket, Twisp, and Winthrop, at the Early Winter Visitor's Center, or by writing the Forest Supervisor, 1240 South Second Street, Okanogan, WA 98840.

Facilities and Services: Campgrounds (some with camper disposal service); picnic areas with grills; boat ramps; swimming beaches; visitor's center.

Special Attraction: Pasayten Wilderness.

Activities: Camping, picnicking, hiking, backpacking, hunting, fishing, boating, swimming, horseback riding, winter sports.

Wildlife: Bald eagle, black bear, mountain lion, Canada lynx, mountain goat.

MUCH OF Washington's Okanogan has a surprisingly manicured, parklike setting with trees scattered here and there over gently rolling meadows. We note "surprisingly" because immediately west of the forest are the rough and tumble mountains of North Cascades National Park. The orderly forest character even extends into the vast Pasayten

Wilderness whose 500,000 acres comprise nearly one-third of the land mass in the national forest.

Despite its gentle terrain, the forest and its special wilderness attraction are designed for the long-distance hiker. You can pick up a leaflet at a ranger station entitled "25- and 50- Mile Hikes in the Okanogan National Forest" and find sufficient long trails to take to last you several summers! You will find solitude along these trails, because most of them are not crowded. There is also good fishing along the way.

The forest does have some short trails that are worth taking, however, including two that are paved and specially designed for the handicapped. One is the Rainy Lake Trail which begins at the Rainy Pass Rest Area on Highway 20 and winds for a mile through alpine country before ending at Rainy Lake. From the observation deck you can enjoy the rugged glacial peaks in the distance as well as fish for cutthroat trout. The other short, paved path, also originating from Highway 20 at Washington Pass, leads to an overlook for one of the best views in the forest. The strikingly beautiful silhouette of Liberty Bell Mountain dominates the scene, but there are also views of Silver Star Mountain and the impressive jagged, mountainous Early Winter Spires.

Most visitors to the forest will eventually drive Highway 20, the scenic North Cascades Highway which connects the communities of Winthrop and Marblemount. For an exceptionally pretty route as well as a thrilling one, take the back road to Harts Pass that leaves Highway 20 west of Winthrop. Timid drivers should avoid it since it is steep and narrow in places. To do justice to the trip to Harts Pass and back, you will need more than half a day. At Dead Horse Point, where there is a salt lick, be alert for mountain goats. Then take the side road to the old mining town of Chancellor and see the evidence of a long-gone mining operation. From the Harts Pass Overlook are perfect vistas of Methow Valley and Goat Wall, the 2,000-foot cliff that stands guard to the Pasayten Wilderness. Harts Pass is adjacent to the wilderness and is a good base for excursions into it.

Another rewarding side trip from Highway 20 is the trail to Cedar Creek Falls, uphill for nearly one and a half miles from the parking lot along the Sandy Butte Cedar Creek Road. There is actually a series of falls where Cedar Creek has relentlessly cut into its granite base. You will enjoy the lovely, brightly colored Indian paintbrushes and lavender penstemons along the trail during summer.

Several segments of the Okanogan lie east of Tonasket and Oroville, small communities along U.S. Highway 97 and the Okanogan River. Roads from either town will lead to the Big Trees Botanical Area where several mammoth conifers tower high above the forest floor. Nearby Lost Lake is a good place to camp or have a picnic. Visitors to this part of the forest in early October will be rewarded by the gorgeous golden needles of the western larch.

Olympic National Forest

Location: 649,979 acres in northwestern Washington, surrounding Olympic National Park. U.S. Highway 101 encircles the forest. Maps and brochures available at the district ranger stations in Hoodsport, Quilcene, Quinault, Shelton, and Forks or by writing the Forest Supervisor, Box 2288, Olympia, WA 98507.

Facilities and Services: Campgrounds (some with camper disposal service); picnic areas with grills; boat ramps; swimming beaches; lodge.

Special Attraction: Quinault Rain Forest.

Activities: Camping, picnicking, hiking, backpacking, hunting, fishing, boating, swimming, nature study.

Wildlife: Bald eagle, black bear, Roosevelt elk, mountain goat.

TWELVE FEET of rainfall each year, warm temperatures, and a humid atmosphere have created a greenhouse effect in some of the valleys in the Olympic Peninsula, the large land mass in northwestern Washington that is surrounded on three sides by saltwater. The results are forests of giant trees and lush vegetation reminiscent of a tropical rain forest. The most popular of these is in the Olympic Forest. It is the Quinault Rain Forest south of Quinault Lake, where 300-foot-tall trees, giant ferns, and a rich layer of wildflowers are common. The forest service has made it easy to explore this rain forest by constructing the Big Tree Grove Nature Trail. The trail head is between U.S. Highway 101 and the Quinault Ranger Station. Douglas firs are the largest and oldest trees in the forest, but mammoth Sitka spruces and western hemlocks are also present. Among the wildflowers that grow in the dense shade of the forest are the Oregon oxalis, with its cloverlike leaves, wild bleeding heart, several kinds of violets, and vanilla leaf, with its leaves cut into peculiar shapes. Even on a clear day, dampness is evident in the air as you hike into the rain forest. The moisture is conducive to the growth of mosses that cling to everything, giving the forest an overall greenish tone. Another easy trail is the one-fourth-mile Pioneers Path Nature Trail which originates from the Klahowya Campground. This trail is adjacent to the Soleduck River several miles northeast of the town of Forks.

Another pleasant place to visit in the forest is Wynoochee Valley, a few miles north of Montesano. Wynoochee Lake is a center for recreation activities, and Coho Campground on the lake's edge is an adequate place to spend the night. It has some facilities for handicapped persons. There is a half-mile nature trail from the campground that provides an introduction to the area, and there is a more strenuous 12-mile trail that encircles the lake. Nine miles north of Wynoochee Lake, reached either by road or trail, is the 40-foot drop of Wynoochee Falls into a deep green pool.

A unique marine environment also exists in the Olympic, at Seal Rock Beach along the shoreline of Puget Sound. Pick up a Seal Rock Beach brochure from a ranger station so you will be able to recognize such things as acorn barnacles, Pacific oysters, purple shore crabs, Dungeness crabs, and five-inch-wide butter clams. Squawking and diving gulls will be your constant companions. If you are lucky you might see a harbor seal paddling offshore.

Before leaving the Olympic, drive or hike to the summit of Mt. Walker for a sterling view of the entire Olympic Peninsula. If by chance the weather is clear, you can see Puget Sound, Seattle, Mt. Baker, Mt. Rainier, and much of the Olympic Range. There is a picnic area on the summit.

Wenatchee National Forest

Location: 2,464,668 acres in central Washington, east of Seattle. U.S. Highways 2 and 97 are the major routes. Maps and brochures available at the district ranger stations in Chelan, Cle Elum, Entiat, Leavenworth, and Naches or by writing the Forest Supervisor, Box 811, Wenatchee, WA 98801.

Facilities and Services: Campgrounds (some with camper disposal service); picnic areas with grills; boat ramps; swimming beaches.

Special Attractions: Glacier Peak Wilderness; Alpine Lakes Wilderness; Goat Rocks Wilderness.

Activities: Camping, picnicking, hiking, backpacking, hunting, fishing, boating, swimming, horseback riding, winter sports.

Wildlife: Bald eagle, black bear, mountain goat, elk.

THE WENATCHEE is a wilderness lover's paradise, for parts of three superior wilderness areas are in the forest. Glacier Peak Wilderness, which spans the crest of the northern Cascade Mountains, is one of them. It is an area of steep-sided valleys between mountain peaks whose upper slopes have active glaciers on them. There are several one-day hikes, as well as many longer ones. One of the easiest trails goes from the Holden Campground, nestled below pines along Railroad Creek, to sparkling Holden Lake.

Another featured wilderness is the Alpine Lakes. There are as many deep blue mountain lakes in this one wilderness as there are anywhere in the forest. Enchantment Lakes east of Mt. Stuart are the most remote and secluded, and for good reason. The steep trail to them from Icicle Creek climbs nearly 5,500 feet and is in exceptionally rough terrain. It probably will take you two days to reach the lakes. Do not attempt it if you are not in tip-top shape.

Goat Rocks Wilderness, the third wilderness in the Wenatchee, is discussed fully in the Gifford Pinchot National Forest entry.

Most of long, narrow, and deep Lake Chelan is in the

Wenatchee, and there are several comfortable campgrounds along the shore, including one only a short hike away from 50-foot Domke Falls. If you take the commercial tour boat that plies the water of Lake Chelan, you will be able to see the falls from the boat. The lake, by the way, is more than 1,500 feet deep.

The forest service has laid out a 24-mile automobile tour that parallels the scenic White River from the west end of Lake Wenatchee to the turbulent, 100-foot White River Falls. There is a campground near the falls. If you do any climbing in the area, be extremely careful because the rocks are treacherously slippery. From the campground you are only 2 miles from the Glacier Peak Wilderness boundary.

One trip in the forest we like is the Bumping River Road which leads south from Highway 410 west of Yakima to Bumping Lake. The route is through breathtaking scenery. You can camp, fish, boat, and participate in other water-based recreation in the lake, but we like to continue on to Deep Creek Campground at the road's end in the heart of no-man's-land. From near the campsite is a rather easy but highly scenic four-and-a-half-mile trail to Tumac Mountain where there are plentiful views into Mt. Rainier National Park and Gifford Pinchot Forest. You are further rewarded along the trail by mountain meadows dominated by bear grass and by beautiful Twin Sisters Lake.

Chugach National Forest

Location: 6,234,241 acres in southern Alaska, from Anchorage and Seward east to Cordova and Katalla, and including islands in the Prince William Sound south to Afognak Island just above Kodiak Island. Routes 1, 9, and 10 provide access to the forest, but remote areas can be reached only by plane, boat, or by hiking. Maps and brochures available at the district ranger stations in Cordova, Seward, and Anchorage, at the Portage Glacier Visitor's Center, or by writing the Forest Supervisor 2221 E. Northern Lights Blvd., Anchorage, AK 99508.

Facilities and Services: Campgrounds; picnic areas with grills; outlying cabins; lodge; boat ramps; visitor's center.

Special Attractions: Portage Glacier Recreation Area; Nellie Juan Wilderness Study Area; Copper River Delta.

Activities: Camping, picnicking, hiking, backpacking, hunting, fishing, boating, swimming.

Wildlife: Bald eagle, trumpeter swan, Canada goose, Alaskan brown bear, black bear, Dall sheep, mountain goat, elk, moose, lynx, wolverine, seal, otter, wolf.

IF YOU decide to visit the two national forests in Alaska, we suggest that you begin with the Chugach Forest at the modern Portage Glacier Visitor's Center at the western end of glacier-formed Portage Lake, about 50 miles southeast of Anchorage. It is a splendid place to get acquainted with the vast Alaskan glaciers, mountains, lakes, and islands that are found in the forest. The gravel ridge that is under the visitor's center is a terminal moraine that was laid down by the glacier when it ended its forward movements in 1893 and began retreating. The Portage Glacier is still moving forward about 15 inches each day, but because the rate of melting is even faster, the glacier is retreating. A short nature trail near the visitor's center provides a firsthand opportunity to examine a terminal moraine. The windblown shrubs that live near the trail have lost all their branches on the wind-facing side. The Glacier View Trail, nearly 1 mile long, is another informative trail and leads toward Byron Glacier. You will marvel at the ice — it appears blue because of the density of the crystals. There are campgrounds, picnic areas, and even a lodge near the visitor's center that serves meals.

The Nellie Juan Wilderness Study Area is a top forest attraction and part of the wild backcountry that lies south of Portage Glacier on the western side of Prince William Sound. The 600,000 acres contain glacier-peaked mountains towering above vast uncut forests. An eerie mist from the sea often hangs over the area. The beaches along the sea are ruggedly beautiful. Inland, through forests of hemlock and spruce, the elevation increases to 5,300 feet, where purple, lavender, and white heather-filled meadows provide a refreshing contrast to the deep green forests. The streams are filled with salmon and fulfill every angler's greatest dream.

Another attraction and fascinating area of the Chugach is the Copper River Delta east of Cordova. Here the Copper River winds through the Chugach Mountains before emptying into the Gulf of Alaska. The Delta is 50 miles wide and is crossed by the Copper River and its many silt-clogged tributaries, forming tidal flats, ponds, sloughs, and gray mud flats. Each habitat has its own unique plant and animal life. Water birds by the thousands congregate as they stop to feed or rest during their migrations. Trumpeter swans, Canada geese, the threatened bald eagle, and a myriad of ducks are some of the birds to behold. If you are patient, you may see the great Alaskan brown bear, the smaller black bear, Dall sheep, moose, mountain goat, lynx, and wolverine. Offshore are the comical seals and sea otters.

There are several conventional campgrounds and picnic areas in the forest, but the ultimate in primitive outdoor living is the solitude from staying in the outlying cabins provided by the forest service. In the most remote regions of the Chugach, reached only by plane, boat, or foot, are cabins that must be reserved in advance from the ranger stations. It will cost you a very small fee to stay in these primitive structures; they are equipped with a wood or oil stove, tables, and bunks. There is no electricity or plumbing. You will need to take your own bedding, utensils, and oil for the stove. You must get your drinking water from nearby lakes and streams and you should boil it before using.

Tongass National Forest

Location: 16 million acres in southeastern Alaska, east of the Gulf of Alaska and extending from Yakutat Bay south to Portland Inlet, surrounding the cities of Juneau, Sitka, Ketchikan, and Petersburg. Maps and brochures available at the district ranger stations in Wrangell, Petersburg, Sitka, Hoonah, Juneau, Craig, Thorne Bay, and Ketchikan, at the Mendenhall Glacier Visitor's Center, or by writing to one of the three Forest Supervisors at Box 309, Petersburg, Alaska 99833, Box 1980, Sitka, Alaska 99835, or Federal Building, Ketchikan, AK 99901.

Facilities and Services: Campgrounds, picnic areas with grills; outlying cabins; boat ramps; visitor's center.

Special Attractions: Mendenhall Glacier; Tracy Arm–Fords Terror Wilderness; Misty Fiords Wilderness.

Activities: Camping, picnicking, hiking, backpacking, hunting, fishing, boating, kayaking.

Wildlife: Bald eagle, trumpeter swan, Alaskan brown bear, black bear, mountain goat, moose, wolverine, otter, sea lion, Sitka black-tailed deer, killer whale, porpoise, wolf.

THE MENDENHALL GLACIER VISITOR'S CENTER, 13 miles north of Juneau, is the ideal place to get your bearings in the mammoth 16-million-acre Tongass National Forest. Although much of the forest is uncharted and unexplored, you can learn about that which is known at the visitor's center. A large relief map helps you get oriented, and films and slide shows in the auditorium describe this part of Alaska's glacier-carved landscape.

The visitor's center is located near the south shore of Mendenhall Lake and directly across from the forest's star attraction—large, active Mendenhall Glacier. Even though the glacier, a part of the vast Juneau Ice Field, is active today, the rate of melting exceeds its forward advance so that the glacier is actually retreating. There are several viewpoints of the glacier from the trails that lead from the visitor's center. The one-and-a-half-mile Moraine Ecology Trail surrounds a glacier-buried forest and passes an old glacial stream channel. The East Glacier Trail, nearly three miles long, goes over a forest that was covered by glacial ice only 100 years ago. If you look across Nugget Creek toward Mt. Bullard, you might even see a mountain goat. Another short trail goes to Steep Creek where salmon can be seen spawning from July through December.

While you are in the Juneau area, drive the 35-mile Glacier Highway from the north end of the city. It will provide a great opportunity to see mountains, glaciers, and picturesque views of the sea. You might want to stop along the highway and hike the 5-mile trail to Salmon Creek or the more difficult 4-mile trail to Goat Mountain.

Near Sitka, at the western side of Baranof Island, are additional developed areas of the Tongass National Forest. There are a couple of comfortable campgrounds at Starrigavan and along Sawmill Creek, as well as some recommended hiking trails. The Indian River Trail follows the river for five miles, ending at a waterfall. You can fish for rainbow and Dolly Varden trout along the way. More untamed and more scenic is the Mt. Verstovia Trail which passes through forests of Sitka spruce and western hemlock.

As you will soon discover, most of the rest of the Tongass is wild, rugged, and undeveloped. Fifty miles southeast of Juneau is another forest special attraction—the startlingly beautiful Tracy Arm–Fords Terror Wilderness. Occupying 653,000 acres between Stephens Passage and the Canadian border, this region contains tidal flats near the sea, colorful meadows, vast forested areas, huge ice fields, and majestic mountains. Penetrating inland are glacier-formed streams called fiords. Tracy Arm, a fiord with navy blue, icy water, extends for 25 miles between sheer cliffs that rise 2,000 feet. Some of the water of the fiord comes from North and South Sawyer, two retreating glaciers. In places, the water in Tracy Arm is more than 1,200 feet deep. Endicott Arm is an even longer fiord and may have icebergs in it. A narrow tributary to Endicott Arm, called Fords Terror, fills a narrow chasm enclosed by sheer cliffs with surging rapids and whirlpools. There are cruise ships that can be taken down Tracy Arm and Endicott Arm.

Largest of all the wildernesses in any national forest is the Tongass's 2,294,343-acre Misty Fiords Wilderness which extends roughly from the Unuk River in the north to Portland Inlet in the south. The area is known for its deep, long, blue fiords that flow between sheer cliffs rising several thousand feet. Near the Unuk River are lava flows and mineral springs. Active glaciers and high, cold water lakes permeate the wilderness. At the northeast corner and straddling the Canadian border are rugged mountains in the 6,000-to-7,500-foot range—Mt. Stoeckl, Mt. Blaine, Mt. Willibert, Mt. John Jay, and Mt. Upshur.

Since wild animals abound here, keep alert for the threatened bald eagle as well as the trumpeter swan, Alaskan brown bear, black bear, moose, mountain goat, wolverine, and the Sitka black-tailed deer. In the fiords are all five species of Pacific salmon—king, coho, chum, sockeye, and pink—and three highly prized trout—rainbow, coastal cutthroat, and Dolly Varden. During certain seasons sea lions, harbor seals, killer whales, and Dall porpoises may be seen in the saltwater bays.

An unusual formation in Misty Fiords is New Eddystone Rock, a 250-foot-tall remains of a once-active volcano whose cone eroded away, leaving only a lava plug. This rock rises abruptly from the water of Behm Canal. Another region of unusual interest lies along the Red River, where the northernmost stand in the world of Pacific silver fir occurs.

The forest service maintains primitive outlying cabins in remote parts of the Tongass. Equipped only with tables, bunks, and a stove, these cabins must be reserved in advance for a small fee from one of the ranger stations.

Much of the Tongass National Forest can be observed from the elegant Alaska ferry lines, which use the Inside Passage as a marine highway from Seattle, Washington, and Prince Rupert, British Columbia, to such Alaskan towns as Ketchikan, Wrangell, Petersburg, Sitka, Juneau, Haines, and Skagway.

PACIFIC SOUTHWEST

The Pacific Southwest region contains all the national forests in the state of California, which boasts more national forests than any other state in the country. They total 17 and are the Angeles, Cleveland, Eldorado, Inyo, Klamath, Lassen, Los Padres, Mendocino, Modoc, Plumas, San Bernardino, Sequoia, Shasta-Trinity, Sierra, Six Rivers, Stanislaus, and Tahoe.

Angeles National Forest

Location: 654,000 acres in southern California, north and east of Los Angeles. Interstates 5, 10, and 15, U.S. Highway 101, and State Routes 2, 14, and 39 serve the area. Maps and brochures available at the district ranger stations

in Flintridge, Glendora, Saugus, San Fernando, and Pearblossom, at the visitor's centers at Crystal Lake and Chilao, or by writing the Forest Supervisor, 150 South Los Robles, Pasadena, CA 91101.

Facilities and Services: Campgrounds (some with camper disposal service); picnic areas with grills; boat ramps; swimming beaches; visitor's centers.

Special Attraction: San Gabriel Wilderness.

Activities: Camping, picnicking, hiking, backpacking, hunting, fishing, boating, swimming, horseback riding, winter sports.

Wildlife: Nelson bighorn sheep, mountain lion, black bear; endangered fish known as the unarmored three-spine stickleback.

THE ANGELES NATIONAL FOREST, situated at the doorstep of Los Angeles, contains nearly 700,000 acres of the rugged San Gabriel Mountains. Because of the forest's nearness to a major metropolitan area, there are naturally many developed recreation sites, and many of the more interesting regions of the forest are accessible from these recreation areas. One of them, the Mt. Baldy Recreation Area, is at the eastern edge of the forest. A good plan is to camp at Manker Flats near the end of the scenic Glendora Ridge Road, as it is a central location to some of the sightseeing in the forest. Limber up by taking the mile-long trail to magnificent San Antonio Falls which noisily and dramatically cascades 500 feet into its catch basin below. There are two recommended trails to choose from but both are into rugged terrain. The Devil's Backbone Trail is one and is a high ridge trail that connects the Mt. Baldy Notch with 10,000-foot Mt. Baldy. The other trail is called the Chapman Trail and begins at Ice House Lodge and twists and turns for five miles to reach Ice House Saddle at the edge of the Cucamonga Wilderness in the adjacent San Bernardino National Forest. As Chapman Trail crosses Cedar Canyon, you will see handsome, spire-tipped incense cedars. You can also detect them in the air through their distinctive cedar scent. On this same excursion you will also see Bear Flat, a mountain meadow teeming with colorful wildflowers, about one and a half miles up Bear Canyon Trail.

Another recreation area that handily serves as a base for a variety of activities is at Chilao, where a visitor's center has a number of exhibits and information about the forest. From the campground there are a number of short, easy loop trails. For the more hardy, we suggest that you take the Devil's Canyon Trail and go into the San Gabriel Wilderness, the star feature of the forest. The wilderness covers roughly 36,000 acres. This affords an excellent chance to glimpse the rare Nelson bighorn sheep, with its magnificent curved horns, or a black bear, or a mountain lion. You will find that the wilderness has rough terrain; it reaches an elevation of 8,200 feet. On the highest ridges, firs and pines predominate; at the lowest, chaparral is dominant.

Another fine forest recreation area with a visitor's center and a large choice of trails is Crystal Lake. The lake itself is good for both boating and swimming. Cabins that you can rent and a small general food and supply store are nearby. They are several pleasant trails less than a mile long that begin at the campground. A scenic two-miler is the loop trail along Soldier Creek.

For something special in the Angeles, try the following: take the twisting trail to the summit of Mt. Baden-Powell. On the north side of the mountain, wander about in a stand of majestic limber pines estimated to be 1,000 to 2,000 years old. The trail to the summit originates from the Angeles Crest Highway (State Route 2). You may wish to hike the route of the historic Mt. Lowe Pacific Electric Railway, which climbed Echo Mountain and part of Mt. Lowe until its last run on December 5, 1937. A guide booklet available at the visitor's center in the forest explains the history of the railway and describes the stops along the hiking trail.

Another recommended stimulating adventure is to hike or ride horseback the 28-mile Gabrielino National Trail which begins at Chantry Flat, climbs to the 4,100-foot Newcombs Pass, and then takes an undulating route to the Arroyo Seco Canyon and the border of the Angeles Forest near Pasadena. Some of the most breathtaking vistas in southern California unfold before you as you hike this trail.

One of the nation's endangered fish, the unarmored three-spine stickleback, lives in a few streams in the forest. While most kinds of three-spine sticklebacks have hard scales, or armor, to protect themselves from predators, the unarmored three-spine stickleback lacks these protective scales. It has become extremely rare in recent years.

Cleveland National Forest

Location: 420,000 acres in southwestern California, between San Diego and Corona. Interstates 8 and 15 and State Routes 74, 76, 78, and 79 are the major roads. Maps and brochures available at the district ranger stations in Alpine, Escondido, and Santa Ana or by writing the Forest Supervisor, 880 Front St., San Diego, CA 92188.

Facilities and Services: Campgrounds (some with camper disposal service); picnic grounds with grills.

Special Attractions: Agua Tibia Wilderness; unusual trees.

Activities: Camping, picnicking, hiking, backpacking, hunting, fishing, horseback riding, winter sports.

Wildlife: Threatened bald eagle; mountain lion, bobcat, ring-tailed cat, peninsular bighorn sheep.

NORTHEAST OF ESCONDIDO is the Palomar district which contains one of the special attractions of this forest, the Agua Tibia Wilderness, occupying 36,000 acres in the extreme northern corner. This mountain wilderness is sliced by dramatically deep canyons with small pools of water. In it is the Pine Flat Trail which we wholeheartedly recommend. It extends for 7 miles through the entire length of the area, beginning from the convenient Dripping Springs Campground, a good base for a campsite. On this trail we saw one of the largest eastern manzanitas we have ever seen, a huge specimen that was as broad as it was tall.

There are also two unusual kinds of cypress tress growing in the Palomar district worth seeking. The rarest of these is the Cuyamaca cypress which grows among dense chaparral thickets on the southwest slopes of Cuyamaca Mountain. The other is the Tecate cypress; there is a pretty grove of them on Guatay Mountain south of the Guatay Campground on U.S. Highway 80. You will need a good tree manual with you to identify these and other woody plants in the forest.

The forest's Descanso district is located about 40 miles east of San Diego and 20 miles north of the Mexican border. It features the Laguna Mountain Recreation Area, an 8,600-acre rolling plateau. On this plateau is an exceptionally large meadow ringed with unusual trees. You will be able to see the Jeffrey pine with its cones sometimes nearly a foot long, as well as the California black oak. If you look closely you can spot the Coulter pine with its large, heavy cones. On the lower slopes you will find an assemblage of shrubs that includes Palmer's ceanothus with its clusters of white flowers during the summer, birch-leaf mountain mahogany, eastern manzanita, and California scrub oak. There are three lakes scattered in the meadow; unfortunately they do not contain fish, because of the large buildup of algae in them. The forested areas on Laguna Mountain harbor an interesting mixture of wildlife, including mountain lions, bobcats, and ring-tailed cats, as well as acorn woodpeckers, western bluebirds, and mountain chickadees. The summit of the mountain provides spectacular views of the desert country to the east. Of historical interest is the El Prado Ranger Station built in 1910 of hand-hewn logs, still with its original fireplace. The large Burnt Rancheria Campground is an excellent place for spending the night on the mountain, because it is in the heart of the better features. From it you can take the short Desert View Nature Trail.

The northern Trabuco district of the forest is northeast of San Juan Capistrano and includes the Santa Ana Mountain Range. It boasts of several scenic canyons including the popular and easily accessible Juan Canyon as well as the remote San Mateo Canyon. We thoroughly enjoyed the nine-mile Chiquito Trail which passes lovely Chiquito Falls.

Eldorado National Forest

Location: 671,000 acres in central California, east of Sacramento. U.S. Highway 50 and State Route 88 serve the forest. Maps and brochures available at the district ranger stations in Pioneer, Georgetown, Fresh Pond, and Placerville, at the Lake Tahoe Visitor's Center, or by writing the Forest Supervisor, 100 Forni Road, Placerville, CA 95667.

Facilities and Services: Campgrounds (some with camper disposal service); picnic areas with grills; boat ramps; swimming beaches.

Special Attractions: Desolation Wilderness Area; Mokelumne Wilderness Area.

Wildlife: Black bear, mountain lion.

THE ELDORADO is in the famous Mother Lode country of California where 50,000 persons rushed in 1849 after James W. Marshall discovered gold. There are still lingering signs of the old days and boom area in the form of abandoned mines, ghost town rubbles, and obliterated tombstones. This area is located between the foothills of the central Sierra Nevadas and the south end of Lake Tahoe. The lake, now one of California's biggest and most popular recreation sites, was discovered by explorer John Fremont in a party guided by Kit Carson. The forest service maintains a visitor's center at the south end with many historic exhibits; rangers are most helpful in answering questions and leading hikes on some of the nearby short nature trails.

Three miles east of the visitor's center, on either side of the Rubicon River's Rockbound Valley, is the Desolation Wilderness, one of the special attraction areas of the forest. The 63,000 acres of this wilderness used to be desolate, but its nearness to popular Lake Tahoe has resulted in wilderness seekers by the thousands in recent years. A portion of the well-known 15-mile Pacific Crest Trail twists through the entire length of the area. We suggest you join it where it originates at Bayview Campground along State Route 89. On the trail you will pass Maggies Peak, the Velma Lakes, climb over Phipps Pass, and hike along the western slope of Rubicon Peak before ending at Sugar Pine Point State Park near Meeks Bay. From the campground at Wrights Lake near the southwestern corner of the wilderness, you can also undertake another trail that climbs over Rockbound Pass to China Flat. To the south is big, blue Lake Aloha.

West of the Desolation Wilderness is the beautiful but highly developed Crystal Basin Recreation Area. If you do not like crowds, you may not enjoy this area, as it is usually heavily populated. You can escape to the lofty granite peaks of the pine- and fir-clad Crystal Range; these peaks tower above some of the bluest lakes in the United States. Among them are Loon Lake, Union Valley Reservoir, and Ice House Reservoir, and all have campgrounds, picnic areas, and boat ramps.

Eldorado's other special attraction is the Mokelumne Wilderness. A large part is in the Eldorado; the remainder is in the Stanislaus National Forest. The wilderness is dominated by 9,332-foot Mokelumne Peak. From the Woods Lake Campground it is only three and a half miles to Fourth of July Lake. Explorers will also find the deep and rugged Mokelumne Canyon which cuts across the wilderness.

Inyo National Forest

Location: 1,800,000 acres on the eastern side of the Sierra Nevadas in California, near Bishop, Mammoth Lakes, Lone Pine, Big Pine, and Independence, and in a small part of Nevada. U.S. Highway 395 and State Routes 168 and 203 are the major highways. Maps and brochures available at the district ranger stations in Mammoth Lakes, Lee Vining, Lone Pine, and Bishop, at the Mammoth Lakes Visitor's Center, or by writing the Forest Supervisor, 873 N. Main St., Bishop, CA 93514.

Facilities and Services: Campgrounds (some with camper disposal service); picnic areas with grills; boat launching areas; visitor's center.

Special Attractions: Largest groves of ancient bristlecone pines in the world; largest foxtail pine in the world; Mt. Whitney; greatest concentration of bighorn sheep in California; earthquake fault; Minarets Wilderness; Golden Trout Wilderness; John Muir Wilderness.

Activities: Camping, hiking, backpacking, hunting, fishing, boating, swimming, horseback riding, nature study, winter sports.

Wildlife: California bighorn sheep, Tule elk, Sierra red fox, rare golden trout.

THE INYO NATIONAL FOREST is full of special attractions. The greatest problem is what to see first in this sprawling forest.

We recommend heading first for the bristlecone pines, a less strenuous venture to see a nearly unbelievable work of nature. A paved road east from the community of Big Pine climbs out of Owens Valley and winds for nearly 20 miles to the Schulman Grove, for a while using the historic old Westgard Pass road. Every bristlecone pine we have ever seen has always lifted our spirits, when we realize that many of these gnarled trees are more than 1,000 years old. The Schulman Grove on the slopes of the White Mountains is particularly inspirational, for this grove contains the greatest concentration of bristlecone pines in the world. We recommend following the trail from the grove parking area, but take your time as you are walking at an altitude of 10,000 feet. After about ¼ mile you will come upon Pine Alpha, a grotesquely gnarled 4,300-year-old tree reputed to be the oldest living thing in the world. A few clusters of green needles are all that indicate the tree is still alive. Some distance along the trail, an ancient tree given the name Methuselah reveals its beauty. Botanists call these trees "living driftwoods" because they have been lashed and pelted by centuries of windblown particles.

Twelve miles up a rather poor road from Schulman Grove, and 1,000 feet higher, is a second group of bristlecone pines. One of these, the Patriarch, has been declared the largest of all ancient trees in the world. It is a hard-to-forget, many-stemmed monarch with a circumference of nearly 38 feet, yet the tree is only about 40 feet tall.

Another part of the Inyo is famous for having the largest foxtail pine in the world. Forest rangers can give directions to this tree. Named for the dense clusters of needles at the end of each branch that resemble the bushy tail of a fox, this giant specimen has a circumference of nearly 23 feet.

Another big attraction in the Inyo is Mt. Whitney. Experienced and hardy hikers can reach its summit at 14,494 feet, which makes it the highest mountain peak in the lower 48 states. The first half mile of the climb to the top is steep and strenuous, but then the going gets easier as you reach Lone Pine Lake, about two and a half miles later. No matter how far one goes on the trail, one is bound to have some breathtaking experiences. Hikers planning to continue to the top should allow two days, camping overnight at Mirror Lake. Precautions should be taken against sudden summer storms. Even if you do not conquer the summit, we suggest driving the narrow, paved mountain road west out of Lone Pine to Whitney Portal. The portal, at an elevation of 8,367 feet, is the usual starting point for hikers to the summit. It has a neat campground and picnic area nestled beneath scented pines that make sleeping come easily. A small store provides supplies to campers, picnickers, and hikers. There is fishing in the pond near the parking lot; one can also watch and listen to Lone Pine Creek as it rushes over granite rocks. Those not tackling the summit can also look in awe at the formidable 12,944-foot Lone Pine Peak to the south. While in this area, which is near the crest of the Sierras, keep on the lookout for rare bighorn sheep, particularly between Mt. Whitney and Convict Lake. These sheep, fully protected by law, are simply gorgeous animals with curved horns that are a perfect work of art. The forest has the largest population of them in the state.

Still another big attraction in the forest is the earthquake fault near the bustling resort area of Mammoth Lakes. You can walk right down into the deep, narrow fissures where an ancient earthquake has split the ground. Try to imagine the violence and rumbling within the earth's interior that caused it. By chance, the day we walked into this chasm and looked up beyond the sheer walls on either side, we had a double experience; we saw our first sun dog, as the so-called phenomenal rings that encircle the sun are labeled. A campground and picnic area are adjacent.

Although wildflowers abound in the Inyo, we recommend the Minaret Vista off State Route 203 about six miles west of Mammoth Lakes. With the jagged peaks of the Minarets as an inspirational backdrop, this meadowlike setting is ablaze with orange, red, and yellow paintbrushes.

For more of the wonders of nature in the Inyo, drive north from Mammoth Lakes along U.S. Highway 395 toward mysterious Mono Lake, with its haunting, gnomelike lime-stone rocks rising out of the water. This area provides a good example of what can happen in aeons of volcanic activity. Extinct craters, glass mountains, and valleys of pumice fill the region. Two of the craters, reached by a back road west off Highway 395, are worth a visit. A half-mile trail from the parking lot leads to them. One has a milky green lake in its bottom; the other has a deep blue lake. Fishermen should forget their gear as both lakes lack fish. The glass mountain, called Obsidian Dome, is a slick and shiny black mountain of volcanic glass, sometimes clear, sometimes opaque. Picking up small pieces of this glassy rock and the nearly weightless pieces of pumice nearby is a pastime that can be pursued for hours. A rather rough, two-and-a-half-mile road west from Highway 395 winds to the parking lot at Obsidian Dome. In driving off the main highway in this region, it is essential that your vehicle stay on the packed-down tracks of the back roads, for a slip into the soft pumice that borders many of these roads will mean getting stuck and will result in all sorts of frustrations.

Three wilderness areas permit backcountry experiences in the Inyo. The Minarets Wilderness contains the rough Minaret Mountains; the rare golden trout inhabits streams in the Golden Trout Wilderness; the John Muir Wilderness has high alpine terrain with fragile, tundralike wildflowers, many of which lie close to the ground.

There are some 100 available campgrounds in the Inyo, which gives you a range of choices. We like the one at Whitney Portal where the sounds of the Lone Pine Creek Cascade can be heard and bring about sleep before we have a chance to savor the experiences of the day. There is a worthwhile visitor's center near Mammoth Lakes with many exhibits and a source of information about the area.

Klamath National Forest

Location: 1,680,000 acres in northern California on either side of Yreka and Weed and extending just over the Oregon border. Interstate 5, U.S. Highway 97, and State Routes 3, 89, and 96 are the major accesses. Maps and brochures available at the district ranger stations in Klamath River, Happy Camp, Sawyers Bar, Mt. Hebron, Somes Bar, and Fort Jones or by writing the Forest Supervisor, 1312 Fairlane Road, Yreka, CA 96097.

Facilities and Services: Campgrounds (some with camper disposal service); picnic areas with grills; boat ramps.

Special Attractions: Marble Mountains Wilderness; Salmon-Trinity Alps Primitive Area.

Activities: Camping, picnicking, hiking, backpacking, hunting, fishing, boating, rafting, nature study, horseback riding.

Wildlife: Threatened bald eagle; black bear, mountain lion, bobcat; rare Siskiyou Mountain salamander.

THE GREAT NATURAL DIVERSITY in the Klamath plays a major role in making this one of the more rewarding national forests in California. Most of the forest is in the Coast Range Mountains and includes the Klamath, Siskiyou, Scott, Salmon, and Marble mountains. To the east, a southern spur of the Cascades extends into the forest, and east of that is the lower Modoc Plateau. The Cascades and Modoc Plateau comprise the Goosenest district of the forest. Much of this district is noted for its geological features. Northwest of Medicine Lake is the Crater Glass Flow, where glasslike rocks that were laid down during volcanic action are found. In the vicinity is a huge lava flow known as the Callahan Lava Flow. Many ice caves and lava tubes may be found here. West of Mt. Hebron is an unusual geological opening known as the Hole-in-Ground, a huge depression in the earth's crust. Most of these features can be reached by short hikes.

In the western segment of the forest, the topography is more rugged, with jagged peaks, deep canyons, and swift-flowing streams contrasting with tranquil alpine lakes. The Marble Mountains Wilderness is a marvelous place to explore the forest. Marble Mountain itself is a dramatic landmark of white and gray marble that contains many fossils of marine organisms. Because of the extreme ruggedness of the area, almost none of the trees in the wilderness has ever been cut. The forest service estimates 187,000 acres of virgin forest here. Extensive stands of Douglas fir dominate the forests at the lower elevations, while a little higher up, Shasta and white firs become common. Around tiny Diamond Lake grows the rare and beautiful silver fir at its southernmost location in the country, while the whitebark and foxtail pines, two trees more common in the southern Sierra Nevadas, grow on Boulder Peak. The forest service map and guide sheet to the Marble Mountains Wilderness have a handy identification key and pictures so that all the conifers of the wilderness can be readily identified.

A very small portion of the Salmon-Trinity Alps Primitive Area, another forest prime attraction, is in the Klamath National Forest. A five-mile hike from the Big Flat Campground to the beautiful Caribou Lakes and back makes a fine one-day outing.

In the area around Seiad Valley, a few miles south of the Oregon border, is the southernmost stand of the Baker cypress, a lovely rare tree that is confined to two counties in southwestern Oregon and three counties in northern California. In the area north of Seiad Valley is the very unusual Siskiyou Mountain salamander, an endangered amphibian in California.

If you want to seek out one of the most secluded lakes in the forest, hike the trail that leads away from the South Fork Salmon River Road near Summerville to Rush Creek Lake. You will have to hike about 15 miles along Rush Creek and McNeil Creek and finally on a ridge trail to the north. After picking your way around some granite boulders, you will drop down to the isolated two-acre Rush Creek Lake. On the north side of the lake is a grove of white fir and western cedars that makes a good place to pitch camp before starting back the next day. The trail is a good introduction to the plant life of the Klamath. Douglas fir, white fir, western cedar, and sugar pine loom above showy patches of columbines, tiger lilies, and penstemons.

Lassen National Forest

Location: 1,060,000 acres in northeastern California, between Redding and Susanville. State Routes 36, 44, and 89 are the major highways in the forest. Maps and brochures available at the district ranger stations in Chester, Susanville, and Fall River Mills, at the Lassen Volcanic National Park Visitor's Center, or by writing the Forest Supervisor, 707 Nevada St., Susanville, CA 96130.

Facilities and Services: Campgrounds (some with camper disposal service); picnic areas with grills; boat ramps; swimming beaches.

Special Attractions: Volcanic phenomena; Caribou Wilderness; Thousand Lakes Wilderness.

Activities: Camping, picnicking, hiking, backpacking, hunting, fishing, boating, swimming, horseback riding, rock hounding, nature study.

Wildlife: Threatened bald eagle; black bear, osprey.

EVIDENCE OF volcanic activity is everywhere in the Lassen, which is located at the northern end of the Sierra Nevadas and the south end of the Cascade Range. The forest completely surrounds Lassen Volcanic National Park, whose Mt. Lassen last erupted in 1915. Signs of this and previous eruptions can be seen in the forest, particularly in the Hat Creek area.

We began our exploration of the volcanic sights and special attractions by taking the Spatter Cone Crest Trail from the amphitheater at Hat Creek Campground. There is an enormous amount of volcanic evidence to be witnessed. To the end of the trail and back is two miles, in which 16 spatter cones and an assortment of lava tubes, blowholes, and volcanic domes can be seen. Spatter cones are the names given to steep-sloped piles of lava that are shaped like volcanoes. The first one seen from the trail has a crater 33 feet in diameter. The spatter cone numbered 15 has a conspicuous opening called a blowhole. Through this blowhole great quantities of steam poured out in the past. Two miles north of all these engrossing sights is the fascinating Subway Cave through which lava flowed less than 2,000 years ago. The cave, which tunnels for 1,300 feet, is entered at the Devil's Doorway and exited at the Rattlesnake Collapse, two ominous-sounding portals. Other marvelous place-names along the way include Stubtoe Hall; Lucifer's Cul-de-sac in which the largest "room" is called the Sancturm; and Lavacicle Lane. The roof height of the cave varies from 4 to 17 feet. Since the cave is in total darkness, it is necessary to have a flashlight with you.

In spite of all the macabre thrills found in the volcano areas, much of the Lassen is gentle mountainous terrain with postcard-pretty lakes and wildflower-laden meadows. This is well exemplified in the Caribou Wilderness, even though it has a high point of 8,374 feet. This wilderness, a forest attraction for visitors, is adjacent to the eastern boundary of Lassen Volcanic National Park and as a consequence it, too, has many reminders of volcanic action. Two of the highest peaks are Black Cinder Cone and Red Cinder. At Cone Lake a trail enters the wilderness and connects with many other trails for pleasant hikes around such beautiful bodies of water as Gem Lake, Emerald Lake, Triangle Lake, and the string of Hidden Lakes.

Another special attraction of the forest is the more rugged Thousand Lakes Wilderness; some peaks rise above the timberline. Magee Peak is a challenge to hikers, with the interesting Gray Cliffs to its north and Red Cliffs to its northwest. An excellent but difficult trail leads to Magee Peak and between the Gray and Red cliffs into the glacier-carved Thousand Lakes Valley. There may not be a thousand lakes in this valley, but there are enough to please lovers of tranquil lakes. Where the trail enters higher elevations, beautiful red fir trees appear. In lower areas white firs and ponderosa pines predominate.

Two large lakes, the natural Eagle Lake and man-made Lake Almanor, are mostly outside the boundaries of the forest, but the forest service maintains recreation areas on some of the shorelines. Both lakes are suitable for boating, swimming, waterskiing, fishing, and other water sports activities. A panoramic view that includes Eagle Lake, Mt. Lassen, and Mt. Shasta can be seen from the solar-operated Antelope Lookout Building a few miles west of Eagle Lake.

At Lassen Volcanic National Park, a visitor's center is operated by the park service and the Lassen National Forest.

Los Padres National Forest

Location: 1,752,000 acres in west-central and southern California in two units, one in the mountains north of Ventura and Santa Barbara and east of San Luis Obispo, the other south of Monterey. U.S. Highway 101 and State Routes 1, 33, and 154 are the major roads. Maps and brochures available at the district ranger stations in King City, Frazier Park, Ojai, Santa Barbara, and Santa Maria or by writing the Forest Supervisor, 42 Aero Camino, Goleta, CA 93117.

Facilities and Services: Campgrounds (some with camper disposal service); picnic areas with grills; swimming beaches.

Special Attractions: San Rafael Wilderness; Santa Lucia Wilderness; Ventana Wilderness; redwood trees.

Activities: Camping, picnicking, hiking, backpacking, hunting, fishing, swimming, horseback riding, nature study.

Wildlife: Endangered or threatened California condor, bald eagle, American peregrine falcon, and San Joaquin kit fox; mountain lion, Mt. Pinos chipmunk, Mt. Pinos blue grouse.

THE MAIN ATTRACTION of the Los Padres are its wonderful wilderness areas.

The San Rafael Wilderness is in the mountains north of Santa Barbara. One of its most popular sights is Hurricane Deck in the heart of the area. This is a region of spectacular cliffs that are pocked with "caves" formed through intensive, almost hurricane-strength wind action. Some of these caves contain Indian writings. There is a good trail to the deck.

Just outside the northeast border of the San Rafael Wilderness is a highly scenic area known as Painted Rocks where strange-shaped rocks and more Indian writings are found. There is an excellent trail to these rocks from Branch Canyon south of New Cuyama, which passes through Lyon Canyon. On the western edge of the wilderness, where the Sisquoc and Manzana creeks come together, is an abandoned 1930s schoolhouse with digger pine sidings and a California sycamore floor.

The nearby and small Santa Lucia Wilderness is an attraction and worth a visit. It is a few miles northeast of San Luis Obispo and is bisected by picturesque Lopez Canyon. A few miles northwest of the wilderness, but still in the Santa Lucia Mountains, is the exciting crest trail from Cuesta Summit to the Cerro Alto Lookout. A must for plant lovers, this trail can be hiked or driven. The rare Sargent cypress forms large, dense groves along the top of the ridge. In addition, small stands of the large-coned Coulter pine are here, along with several rare wildflowers.

South of Monterey and extending from the coast to the Santa Lucia Mountains is the scenic Monterey district of the Los Padres. The Ventana Wilderness, another wilderness of note, is at the north end of this unit in an area of extremely steep slopes, sharp ridges, and V-shaped valleys. Elevations range from 600 feet along the Big Sur River to 5,750 feet. Both the geology and botany are superior. The Ventana (Spanish for "window") Double Cone has a windowlike notch in the ridge just west of the peak. Growing along some of the upper streams is bristlecone fir, one of the rarest trees in the country, which lives in an area only 58 miles long and 12 miles wide. There are also stands of the coast redwood, as well as eight or nine wildflowers on California's rare plant list. Bring along a guide to identify the latter.

The broad summit of 8,831-foot Mt. Pinos is a good place to view the easternmost portion of the Los Padres. It is also the best place to see the highly endangered California condor soaring overhead. There probably are only about 30 of these birds with the 9-foot wing spread left, and they all nest in a restricted area in the forest. Mt. Pinos is also the only home in the world for the Mr. Pinos chipmunk and the Mt. Pinos blue grouse. The trees near the barren summit are mostly limber pines. Farther downslope is an extensive forest of white firs and Jeffrey pines.

A five-mile trail connects Mt. Pinos to Sawmill Mountain and Mt. Cerro Noreste. In the Camp Alto Campground on Mt. Cerro Noreste is a large Jeffrey pine with a cross emblazoned on it by Spanish sheepherders during the 1850s. (There is also an automobile road that comes into the campground from the west.)

East of Mt. Pinos is the unusual big-cone Douglas fir, a species with cones much larger than the regular Douglas fir. One specimen with a trunk measuring 24 feet in circumference is in Pleito Canyon. Ask the district ranger for directions to it.

In the vast area between Mt. Pinos and the community of Ojai are three special points of interest. Thorn Point has a lookout on its 6,935-foot summit that provides views of the Pacific Ocean and some of the Channel Islands on a clear day. The point is reached by a steep three-mile hike from the lovely and secluded Thorn Meadows Campground. Near the Ozona Picnic Ground, where the forest road crosses Reyes Creek, is the old Reyes Adobe House that was built in 1850 by one of the original homesteaders in the valley.

Two of the rooms in the adobe are original. For interesting exploring, hike the intertwining trails around Piedra Blanca where there is a remarkable assemblage of white rocks that rise 100 feet into the air in an area covering 600 acres. To the west, in the mountains above Santa Barbara, are the Caliente Hot Springs. The 118-degree water is piped into a tank where bathing in the warm water is popular.

The southermost grove of coast redwoods in the world is in Redwood Gulch off scenic Highway 1. Also accessible by short hikes off Highway 1 are Jade Cove, where rock hounds can search for native jade (it is mostly gray in this area), two ocean beaches at Sand Dollar and Pfeiffer Beach (swimming not recommended because of the currents), and Salmon Creek Falls. The falls are along the beautiful Salmon Creek Trail where there is also the southermost grove of Douglas firs along the California coast.

Finally, if you should be hiking in the valley foothills near the Tehachapi Mountains and see a small, slender fox, it will likely be the endangered San Joaquin kit fox.

Mendocino National Forest

Location: 882,000 acres in northwestern California, between Redding and Ukiah. State Routes 20 and 36 are near the south and north boundaries. Maps and brochures available at the district ranger stations in Corning, Covelo, Stonyford, and Upper Lake or by writing the Forest Supervisor, 420 E. Laurel St., Willows, CA 95988.

Facilities and Services: Campgrounds (some with camper disposal service); picnic areas with grills; boat ramps; swimming beaches.

Special Attractions: Yolla Bolly–Middle Eel Wilderness; unusual botanical area.

Activities: Camping, picnicking, hiking, backpacking, hunting, fishing, rock hounding, boating, swimming, horseback riding, nature study, winter sports, hang gliding.

Wildlife: Threatened bald eagle; bear, bobcat, mountain lion.

DO NOT BE SURPRISED to look up in the sky in the Mendocino Forest and see human daredevils gliding gracefully through the air. This is because hang gliding is popular in this forest. The truly adventurous hang glider drops 6,000 feet from Hull Mountain to the Gravelly Valley Airstrip. Shorter drops depart from Elk Mountain and Tool Cache Ridge.

For those of us preferring to keep our feet on the ground, there are hiking trails galore in the Mendocino. Many are in the Yolla Bolly–Middle Eel Wilderness, one of the star attractions of the forest located in the north Coast Range Mountains. (One-third of this 111,000-acre wilderness is in the Trinity National Forest.) The wilderness is named for two Yolla Bolly mountains and the Middle Fork of the Eel River. (Yolla Bolly are Indian words meaning "snow-covered high peaks.") One good loop hike departs from George's Valley Trail Head and follows Wrights Ridge past Windy Mountain, then joins the Soldier Ridge Trail which passes between Sugarloaf Mountain and Solomon Peak before returning to the trail head. At the higher elevations, the trail goes beneath ponderosa and sugar pines, red, white, and Douglas firs, and incense cedars. Just after entering the wilderness, the trail crosses Balm of Gilead Creek where fishing for rainbow trout is usually exceptionally good. Other trails in the wilderness lead to Horsehead Mountain, Hammerhorn Mountain, and the highest of them all, 8,092-foot South Yolla Bolly Mountain. Beware as you walk, as rangers warn of many rattlesnakes in the area.

Near the wilderness the three-mile Pack Saddle Trail and the four-mile Lake Shore Trail, both moderately easy, are around parts of Lake Pillsbury. The Snow Mountain Area, which encompasses the gorge of the Middle Fork of Stony Creek, also has a good trail system. You can also view the flattopped, often barren summits of East and West Snow Mountains.

On the eastern edge of the North Coast Range Mountains, in an area of rugged ridges and narrow valleys, is one of the most fascinating botanical areas—and a special attraction in the forest. Growing in rocky terrain near Frenzel Creek is the northernmost stand in the world of Sargent cypress. The largest of this uncommon tree in this stand is about 70 feet tall and 2 feet in diameter. Growing nearby is a smaller-statured plant known as the MacNab cypress. A good tree guide will help you distinguish these trees.

Rock hounds will enjoy hunting for agate and jasper along Big Stony Creek and for jadeite and nephrite along the Eel River. Obtain specific localities for these rocks from the district rangers.

Modoc National Forest

Location: 1,654,032 acres in northeastern California, on either side of Alturas. U.S. Highway 395 and State Routes 139 and 299 are the major access roads. Maps and brochures available at the district ranger stations in Cedarville, Adin, Canby, and Tulelake or by writing the Forest Supervisor, 441 N. Main St., Alturas, CA 96101.

Facilities and Services: Campgrounds (some with camper disposal service); picnic areas with grills; boat ramp; swimming beaches.

Special Attractions: South Warner Wilderness; Glass Mountain and other volcanic features.

Activities: Camping, picnicking, hiking, backpacking, hunting, fishing, boating, swimming, horseback riding, rock hounding.

Wildlife: Bear, mountain lion, golden eagle.

THE NORTHEASTERN CORNER of California is far less visited than other parts of the state; it is not because of a lack of things to see and do, for the Modoc Forest has a wealth of entertaining features. We recommend starting at the Warner Mountains, an isolated range east of Alturas about 80 miles long and 10 miles wide. In the southern wild end of the range is the 69,540-acre South Warner Wilderness, a region of alpine lakes, high peaks, clear streams, rugged canyons, and grassy meadows. Warren Peak, whose summit is dramatically reflected on Patterson Lake, is a challenging mountain. Hikers have the option of taking its exciting Summit Trail, but the trail is most difficult, as it covers 27 miles of rugged but magnificent alpine scenery stretching from

Porter Reservoir on the north to Patterson Campground on the south. En route it passes three major mountains, Squaw Peak, Warren Peak, and Eagle Peak, as well as unusual geological formations called Devil's Knob and the Slide. Easier and shorter forays into the wilderness lead to Mill Creek Falls and Clear Lake from the Mill Creek Falls Campground, and to a natural stone bridge from the Emerson Campground. We found the Mill Creek Falls Campground one of the most pleasant in the forest.

By contrast, if you visit the more settled northern half of the Warner Mountains you will see remains of civilization, including old gold and silver camps like the ones on Mt. Vida, and you will see beautiful views, such as that from the top of Sugar Hill. There is also a sad historic site at the scenic mountain crossing called Fandango Pass. Here a large party of emigrants mistook Goose Lake for a part of the Pacific Ocean and made camp. As they celebrated and danced the fandango, a band of Indians silently encircled the campsite and killed all but two of the pioneers.

The larger part of the Modoc lies west of Alturas and is generally an area of flat, rocky plateaus and steep mountains. Southwest of Goose Lake is the plateau country. Western juniper dominates the landscape. In fact, at Devil's Garden, it has its finest display in an 800-acre area that the forest service has designated the Devil's Garden Natural Area. To get to this site, drive the rough road through Rimrock Valley.

On the western side of the forest, where volcanic action has played the dominant role in shaping the terrain, are the fascinating Medicine Lake Highlands. Clear and beautiful Medicine Lake, with no known outlets, and with depths well over 100 feet, is the central attraction of the area. Most unusual, however, is Glass Mountain, the result of a tremendous volcanic obsidian flow, which covers close to 4,000 acres. Multicolored, glasslike obsidian rock is found on the eastern face of the mountain. There is another glass flow less than a mile north of Medicine Lake, but this one, unfortunately, is composed of dull gray rocks. Six miles southeast of the lake is an area where black lava chunks are found throughout a forest that apparently has never been cut. This 14-square-mile area is known as the Burnt Lava Flow Virgin Area. A good road circles the northern half, with a stop at an ice cave that was formed as a result of a collapsed lava tube, a tunnellike structure through which lava flowed during a past volcanic eruption.

Plumas National Forest

Location: 1,163,657 acres in northern California, around Quincy. U.S. Highway 395 and State Routes 70 and 89 are the major highways. Maps and brochures available at the district ranger stations in Blairsden, Greenville, Challenge, Oroville, Milford, and Quincy or by writing the Forest Supervisor, 159 Lawrence St., Quincy, CA 95971.

Facilities and Services: Campgrounds (some with camper disposal service); picnic areas with grills; boat ramps; swimming beaches.

Special Attractions: Feather Falls Scenic Area; Middle Fork of the Feather Wild and Scenic River; Butterfly Valley Botanical Area.

Activities: Camping, picnicking, hiking, backpacking, hunting, fishing, boating, swimming, waterskiing, horseback riding, nature study, winter sports.

Wildlife: Black bear, osprey; threatened bald eagle.

WE DECIDED to view one of the great special attractions of this forest on a recent trip. We left Oroville and headed northeast to the parking area a short distance north of Feather Falls Village. We were about to see Feather Falls, reported to be the sixth highest waterfall in the United States. We started the trail by zigzagging downhill for nearly a mile, wincing slightly at the thought of having to climb up this steep hill on our way back! Our excitement intensified as we made our way along the 3½-mile trail toward the falls, pausing occasionally to admire a flowing mountain stream or to watch a small, bubbling spring. We could hear the thunder of the falls long before we could see them. We finally got to the rocky pinnacle where the forest service has built a sturdy observation deck. Across from us, where the Fall River comes to the edge of a granite precipice, was Feather Falls, all 640 feet of it plunging before rushing off into the Middle Fork of the Feather River. The falls are the center of attraction of the Feather Falls Scenic Area, a region surrounding the Bald Rock Canyon section of the Middle Fork of the Feather River. We had not expected so many wonderful sights. Among our discoveries was Bald Rock Dome, with its bare, rounded, granite summit on the west side of the river. It should not be missed. A twisting trail off of the Bear Creek Road leads to this landmark. A short distance beyond the dome is another beautiful cascading waterfall along the Middle Fork. The entire 108 miles of the Middle Fork of the Feather River is a designated Wild and Scenic River. Some sections, such as Bald Rock Canyon and Upper Canyon, are so untouched that it is virtually impossible to hike all the way through them.

If more than one day can be spent in the Feather Falls Scenic Area, plan to stay overnight at the remote and peaceful Milsap Bar Campground where the South Branch joins Middle Fork. A three-mile hike along the South Branch leads to the South Branch Falls, a different kind of waterfall where there are nine separate falls plummeting one after another over the rocky domes until the last clear pool is reached. The drops range from 30 to 150 feet. It is a fantastic sight.

Less isolated in the forest is the Antelope Lake Recreation Area where camping, picnicking, fishing, boating, and swimming are all possible. There is a nature trail near the Lone Rock Campground, and naturalists' programs are frequently presented in the amphitheater. Near the east end of Antelope Lake is a log cabin where a rancher and his family lived long ago. Two children's graves are on a small hill above the cabin.

Another popular spot is Frenchman Reservoir, with plenty of water-related activities available. The area is particularly attractive because the Little Lost Chance Canyon below the dam has dramatic sheer canyon walls. From Bucks Lake, at the Bucks Creek Recreation Area, is an interesting 7-mile trail to Bald Mountain Lookout where superb views of the forest are in every direction.

An article in *Fremontia*, the journal of the California Native Plant Society, enticed us to seek Butterfly Valley, another star attraction of the forest northwest of Quincy, where the plant life is so unique that the forest service has designated the region as a Botanical Area. There is a network of seeps and bogs in the open area, and many rare plants find their homes in these wet places. Most notable is the insectivorous cobra plant. It has an arching hood with a forked projection that has a fanciful resemblance to a deadly cobra snake. Three other insectivorous plants have been found in the valley, but keep in mind that all plants in the Botanical Area are protected by law. Butterfly Valley itself is serene and peaceful, an open, parklike area rimmed by a forest of conifers.

San Bernardino National Forest

Location: 810,000 acres in southern California, north of San Bernardino and Riverside. Interstates 10 and 15 and State Routes 18, 38, and 74 are the major roads. Maps and brochures available at the district ranger stations in Rimforest, Fawnskin, Fontana, Mentone, and Idyllwild or by writing the Forest Supervisor, 144 N. Mountain View, San Bernardino, CA 92408.

Facilities and Services: Campgrounds (some with camper disposal service); picnic areas with grills; swimming beach.

Special Attractions: San Gorgonio Wilderness; Cucamonga Wilderness; San Jacinto Wilderness.

Activities: Camping, picnicking, hiking, backpacking, hunting, fishing, swimming, nature study, horseback riding, winter sports.

Wildlife: Peninsular bighorn sheep; endangered or threatened bald eagle; golden eagle, acorn woodpecker, great horned owl; endangered fish known as the unarmored threespine stickleback.

LONG AFTER a visit to the San Bernardino, most visitors remember the distinctive form of an arrowhead that rears up from the mountains behind the city of San Bernardino. The Arrowhead, as it is called, is a landmark; its conspicuous appearance is the result of the light green sage and lupine that grow in the shape of a perfect arrowhead, surrounded by contrasting dark green conifers.

The forest is characterized by sharp ridges and rugged canyons, all a part of four mountain ranges—San Bernardino, San Gabriel, San Jacinto, Santa Rosa. Most of the highest points on these mountain plateaus seldom exceed a 7,000-foot elevation. Peninsular bighorn sheep are occasionally seen in the San Gorgonio Mountain area, and birds such as the acorn woodpecker, great horned owl, golden eagle, and evening grosbeak may be observed.

The forest is known for its wilderness areas. One is the San Gorgonio Wilderness. It is located south of Big Bear Lake, and centered around San Gorgonio Mountain, with its steely gray granite peak. The area is rich in scenery, with gorgeous high mountain meadows, steep slopes with awesome boulders, glacier-carved basins, and bubbling streams. Twisted limber pines add an artistic touch to the scene. Just outside the south boundary is 480-foot Big Falls which makes two spectacular drops during its descent.

Nearby is the famous Cucamonga Wilderness. Our favorite hike into this wilderness is from the Joe Elliott Big Tree Campground where there is a giant specimen of sugar pine; the tree has cones up to 18 inches long. By means of a steep and rugged route, the Wilderness Crest Trail goes to Cucamonga Peak where views of both the desert and the Pacific Ocean can be enjoyed on a clear day.

Near the picturesque mountain community of Idyllwild is the San Jacinto Wilderness, with some of the most challenging trails in the forest. The one from Chinquapin Flats to Tahquitz Peak offers spectacular vistas, particularly from the lookout on the peak. To the west is prominent Lily Rock.

Southwest of Cajon Summit and reached by a half-mile trail from State Route 38 is a strange group of tilted rocks that are literally full of holes of various sizes. Known as Mormon Rocks, they are located along the San Andreas Fault and about six miles from the Cucamonga Wilderness.

The 100-mile, scenic, loop Rim-of-the-World Highway provides a good cross section of the forest and also serves as a "jumping off" place for many of the trails. A sweeping view toward Big Bear Lake from Lakeview Point is worth a stop.

For those interested in history, a drive or hike along the Gold Fever Loop Trail passes through an area of old mining camps and will be sure to stir memories from a bygone era. Beginning at the Big Bear Ranger Station north of Big Bear Lake, the route enters Holcomb Valley and passes the Last Chance Placer Mine and the still-standing log Two Gun Bill's Saloon before coming to a picturesque juniper that served as a hangman's tree. A short distance beyond is the lone remaining cabin of the 1860 mining town of Belleville, an old miner's grave that is surrounded by a hand-carved picket fence, the tiny "pygmy cabin" that served as a barbershop, and old Metzger Mine.

A few miles east of the Gold Fever Trail, in the Horsethief and Cactus Flats area, is a scattering of the strange Joshua trees growing among cacti and other desert vegetation.

Sequoia National Forest

Location: 1,125,000 acres in south-central California, east of Fresno and Bakersfield. State Routes 155, 178, 180, and 190 are the main roads to the forest. Maps and brochures available at the district ranger stations in Clingan's Junction, Springville, California Hot Springs, Bakersfield, and Kernville or by writing the Forest Supervisor, 900 W. Grand Ave., Porterville, CA 93257.

Facilities and Services: Campgrounds (some with camper disposal service); picnic areas with grills; boat ramps; swimming beaches.

Special Attractions: Giant sequoia groves; Dome Wilderness; Golden Trout Wilderness; High Sierra Primitive Area.

Activities: Camping, picnicking, hiking, backpacking, hunting, fishing, boating, swimming, horseback riding, nature study, winter sports.

Wildlife: Endangered or threatened bald eagle, American peregrine falcon, Little Kern golden trout, and blunt-nosed leopard lizard; bighorn sheep.

THE FIRST THOUGHTS of most visitors to the Sequoia National Forest are the sequoia trees, one of the forest's best-known attractions. Let us start with the giants, then. These trees are the most massive trees on earth and are limited to a 260-mile belt on the western slope of the Sierra Nevadas. The forest has 30 groves of them. Some are in pristine conditions; others show the result of wasteful logging practices. The largest specimen in a national forest and the third largest giant sequoia in the world is the Boole Tree in the Converse Mountain area. This tree stands 269 feet tall and has a circumference of 90 feet and a diameter of nearly 36 feet at the base. There is a trail to this tree that passes — sadly — evidence of wanton destruction. More than 8,000 sequoias were cut in the Converse Basin; many fallen ones were never removed because they shattered on their impact with the ground. It is indeed wasteful.

We suggest you hike to the Chicago Stump located near the south end of Converse Basin. This stump may have been from the largest sequoia ever grown. The tree was cut in 1893, chopped into pieces, and sent to the Chicago World's Fair where it was reassembled. Fairgoers, disbelieving a tree could be so big, called the exhibit the California Hoax.

Fortunately, there is still a fine grove of sequoias east of Hume Lake and easily reached by a short trail. Known as the Evans Grove, it consists of about 500 mature trees. As in all sequoia groves, large trees of other species are also present. Huge sugar pines, ponderosa pines, white firs, and incense cedars add to the majesty of each grove. Finally, the southernmost grove anywhere of giant sequoias, the Deer Creek Grove, is in the Sequoia National Forest east of California Hot Springs. This grove of about 30 mature specimens is on a steep slope above the Deer Creek Mill Campground. A good trail passes by all of the large trees here. Camping below this grove is a thrill not soon forgotten.

A rare tree to seek in the Sequoia is the Paiute cypress. Perhaps the finest grove of this gray-leaved cypress tree anywhere is on the north slope of Bald Eagle, easily accessible from Forest Road 27502. If you look around in the forest, you can also find foxtail pines, so named because of their bushy foliage resembling a fox's tail, and digger pines, outstanding for their large, heavy cones.

After seeing the star trees, head for the Dome Wilderness, best reached from the Long Valley Campground on its eastern side. En route, visit Grizzly Falls and Salmon Creek

Falls, two superb waterfalls. A short path from State Route 180 leads to the Grizzly, while a rough trail nearly a mile long goes to the other. West of Grizzly Falls and situated along the South Fork of the Kings River is the large, natural Boyden Cave. Finally you will come upon spectacular barren domes of granite that give the Dome Wilderness its name. They thrust into the air and are conspicuous because of their seemingly bare summits. The easily reached summit of Dome Rock on Forest Highway 90 is a great place to view the surrounding forest. White Dome, Church Dome, and Bart Dome are only three impressive domes in the wilderness; unfortunately they are not easy to get to, because the rough terrain prevents an extensive trail system.

The Golden Trout Wilderness, another sight not to be missed, is largely in the neighboring Inyo Forest. This wild terrain is penetrated by the Kern River and takes its name from the striking golden trout and Little Kern golden trout found in the river. The latter species is on the federal threatened list and fishing for them is forbidden. Mt. Florence, at 12,432 feet, regally towers over the wilderness. Look for the bushy foxtail pine near timberline.

The High Sierra Primitive Area, another site to visit, is adjacent to Kings Canyon National Park and is partly in the Sierra Forest. It is both wild and beautiful, with colorful mountain meadows of wildflowers mixed in among the coniferous forests.

Shasta-Trinity National Forest

Location: 2,153,544 acres in northern California, west of Redding. Interstate 5 and State Routes 3, 36, and 299 are the major roads. Maps and brochures available at the district ranger stations in Platina, Hayfork, Big Bar, Weaverville, Redding, Mt. Shasta, and McCloud, at the Whiskeytown-Shasta-Trinity Visitor's Centers, or by writing the Forest Supervisor, 2400 Washington Ave., Redding, CA 96001.

Facilities and Services: Campgrounds (some with camper disposal service); picnic areas with grills; boat ramps; swimming beaches.

Special Attractions: Mt. Shasta; Yolla Bolly–Middle Eel Wilderness; Salmon-Trinity Alpine Primitive Area; Shasta Recreation Area; Trinity Recreation Area.

Activities: Camping, picnicking, hiking, backpacking, hunting, fishing, boating, swimming, waterskiing, horseback riding.

Wildlife: Black bear, mountain lion, bobcat, golden eagle.

THE SHASTA-TRINITY FOREST sprawls over more than two million acres of dense forests, mountain meadows, deep canyons, and granite peaks, capped by mighty Mt. Shasta at the eastern edge of the forest. Mt. Shasta towers to a height of 14,161 feet and can be seen for dozens of miles. To flatlanders like us, the mountain looks unconquerable, but it can be climbed if you have had some climbing experience, are careful, and are in good physical condition. There are two major climbing routes to the summit. We were told that the route from the Ski Shasta Lodge to the top and back could be accomplished in one day. If you can make it to the top, you will see and smell sulphur fumes seething from fumaroles, a reminder that you are standing on a volcano.

One trail we enjoyed was into narrow Devil's Canyon. The trail begins about four miles from Denny, an old mining town. After crossing a suspension bridge over Devil's Gulch (you can wade across if the bridge looks too scary), you are on your way along the steep-sided canyon. The fascinating feature of this trail is the dense cover of mosses on the rocks, on the ground, on Douglas firs, and even on the tan oaks. We hiked the trail for about three miles before turning back, but the trail continues for another seven miles before it climbs out of the canyon to connect with another trail that is heading for the 7,000-foot Thurston Peaks.

Some of the most remote regions of the forest are in the Yolla Bolly–Middle Eel Wilderness; others are in the Salmon-Trinity Alps Primitive Area. The Yolla Bolly–Middle Eel Wilderness, partly in the Mendocino Forest, straddles the crest of the North Coast Range Mountains. The North Yolla Bolly Mountain is the major peak in the Shasta-Trinity side of the wilderness. Loop trails beginning at Stuart's Gap or Rat Trap Gap trail heads circle this mountain and the adjacent Beegum Basin. The trail passes several grassy openings that the local people refer to as glades.

The Salmon-Trinity Alps Primitive Area is 223,000 acres of sheer mountain beauty. Rugged steep ridges are separated by glacial-cut canyons. Alpine lakes, many of them reflecting adjacent mountain peaks, are abundant. The bonsai forms of stunted conifers add a touch of stark beauty to the higher elevations. Several trails to this area depart from the large Clair Engle Lake. One of the best, which can be done easily in a day, is the Swift Creek Trail which leads to the mountain-enclosed Swift Creek Gorge.

In the nonwilderness sections of the forest, we like the natural bridge that has been carved from the limestone along Hayfork Creek.

Much of the activity in the forest centers around the Shasta and Trinity units of the Whiskeytown-Shasta-Trinity recreation areas. The Shasta section, which encompasses huge Shasta Lake, and the Trinity section, surrounding Clair Engle Lake, are equipped with campgrounds, picnic areas, boat ramps, visitor's centers, nature trails, and amphitheaters.

Sierra National Forest

Location: 1,300,000 acres in central California, east of Merced and Fresno and south of Yosemite National Park. State Routes 41 and 168 are the main accesses. Maps and brochures available at the district ranger stations in Oakhurst, Sanger, Mariposa, North Fork, and Shaver Lake or by writing the Forest Supervisor, 1130 "O" St., Fresno, CA 93721.

Facilities and Services: Campgrounds (some with camper disposal service); picnic areas with grills; boat ramps; swimming beaches.

Special Attractions: Groves of giant sequoia; John Muir Wilderness; Minarets Wilderness; Kaiser Wilderness.

Activities: Camping, picnicking, hiking, backpacking, hunting, fishing, boating, sailing, swimming, waterskiing, horseback riding, nature study, winter sports.

Wildlife: Endangered or threatened bald eagle and Lahontan cutthroat trout; black bear, wolverine.

LIKE THE SEQUOIA FOREST, one of the big attractions of the Sierra is its sequoia trees. The first grove of giant sequoias we had ever visited when no one else was around was at Nelder Grove in this forest. We have never been the same since. These majestic giant forms of life created in us a new sense of reverence and inner peace. This isolated grove of sequoias, bisected by Nelder Creek, is a few miles east of State Route 41 via a narrow dirt road. As we walked the Shadow of the Giants Trail, we looked in virtual disbelief at the towering trees—not only sequoias but immense white firs, sugar pines, and incense cedars, as well. The largest sequoia we saw in this grove rose to 295 feet with a basal circumference of 72 feet. The impact of this forest of mammoths was so great that we scarcely noticed the smaller Pacific dogwoods, white alders, chinquapins, and azaleas growing down at our level.

The other sequoia grove in the Sierra is the McKinley Grove southeast of the Dinkey Creek area. Although the giants here may even be a little more impressive, the grove is without trails and hiking must be made through a dense undergrowth of vegetation and fallen logs.

In addition to the sequoia sites, there are three wilderness areas in the forest we suggest visiting. One is the gigantic John Muir Wilderness, with part of its 500,000 acres in the Inyo National Forest. The wilderness follows the crest of the High Sierra from Mammoth Lakes to Kings Canyon, and then curves around to the Mt. Whitney area. Many of the high mountains are snowcapped year-round. When the snow does melt, bare granite peaks through. Several peaks exceed 12,000 feet and are good places to see glaciers. Lodgepole pine, western white pine, whitebark pine, mountain hemlock, and red fir grow in the areas below timberline. In addition to the scenic high peaks, other features that appealed to us were the brilliant red Vermilion Cliffs, the curious rock formation called the Devil's Bathtub, and beautiful wildflower-filled Graveyard Meadow, all reached from the Vermilion Campground in half a day or so.

The second wilderness, also shared with the Inyo, is the Minarets. The Ritter Range in this wilderness includes Mt.

Davis, Banner Peak, Mt. Ritter, and the Minarets, all above 12,000 feet and all straddling the boundary between the two forests. Lake Catherine and the Twin Island Lakes to the west of this range are strikingly blue and beautiful. The well-known Pacific Crest Trail follows along much of the ridge line in both wildernesses.

The Kaiser Wilderness—another place to plan a visit— is north of Huntington Lake and is centered around Kaiser Peak and its numerous jewellike lakes.

As for camping, there are many lakes and reservoirs in the Sierra that have pleasant campsites about them and that offer great opportunities for boating, sailing, waterskiing, swimming, and fishing. The most popular ones are Huntington Lake, Bass Lake, Shaver Lake, Redinger Lake, and the Pine Flat Reservoir.

During May and June one of the most beautiful as well as one of the rarest shrubs in the country blooms in a small area in the Sierra. The plant is carpenteria, a several-stemmed shrub growing 8 to 16 feet tall and having large white flowers with dense clusters of yellow, pollen-producing stamens in the center. Discovered by explorer John C. Fremont in 1845, this plant is found only in chaparral in an area roughly 17 by 12 miles. One stand has been preserved in the special Carpenteria Botanical Area in the Big Sandy Bluffs region along State Route 168 southeast of Shaver Lake. We recommend a visit when it is in bloom.

Six Rivers National Forest

Location: 980,000 acres in northwestern California, east of Crescent City and Eureka. U.S. Highway 199 and State Routes 36, 96, and 299 are the major roads. Maps and brochures available at the district ranger stations in Gasquet, Willow Creek, Bridgeville, and Orleans or by writing the Forest Supervisor, 507 F St., Eureka, CA 95501.

Facilities and Services: Campgrounds (some with camper disposal service); picnic areas with grills; boat ramp.

Special Attractions: Giant redwoods and other unusual plants.

Activities: Camping, picnicking, hiking, backpacking, hunting, fishing, boating, horseback riding, nature study, river rafting.

Wildlife: Endangered or threatened bald eagle and American peregrine falcon; black bear, elk, wolverine, mountain lion, bobcat.

THE FOREST'S NAME comes from the six roaring rivers that cut across the steep mountains of the North Crest, Salmon, and Siskiyou ranges in the northwest corner of California. The six—the Smith, Klamath, Mad, Van Duzen, Trinity, and Eel—team with two super sport fish, the salmon and steelhead, making this one of the most famous fishing areas in the state of California.

The other star attractions of the forest are botanical in nature. One is the exceptional isolated pockets of the giant redwood. These trees, rising higher than any other living organism, live in dense woods shared by gorgeous rose-colored rhododendrons and a myriad of large ferns. Hiking in a redwood grove gives one the feeling of being in another world.

The other botanical discovery to be savored is the forest's rare and unusual plant life. One pristine backcountry area that is geologically interesting and botanically significant is Bear Basin Butte in the western Siskiyou Mountains, just east of scenic Siskiyou Pass. Brewer's weeping spruce, one of the rarest conifers in the United States, grows on cool north slopes in the Siskiyous, as well as in the Trinity Mountains. This particular stand is among the finest you will ever see. Another rare conifer is the Alaska cedar, a tree usually found in Alaska and western Canada, but you can see it here in the Siskiyou Mountains, its southernmost habitat. On the drier slopes are dense mountain chaparral. Among the common huckleberry oaks, canyon live oaks, and green-leaf manzanitas is Sadler's oak, another plant with a small geographic range. There is also an abundance of wildflowers in the Bear Basin Butte, but we suggest that you take a wildflower field guide with you since many of the lovely plants will be unfamiliar, as they only grow here.

About five miles southeast of Bear Basin, in an untouched and primitive setting, is a beautiful, almost paintinglike scene not to be missed. It is the sparkling, jewellike Island

Lake which lies in a basin below Jedediah Mountain. The slopes from the crest of the mountain to the lake have a sparse forest of deep green conifers.

Since the Six Rivers National Forest has a total north-to-south distance of about 135 miles, there is a great diversity of environmental conditions in the forest. In the lower areas along streams, alders, willows, western dogwoods, and big-leaf and vine maples are frequently noted. The yellow and red autumnal foliage of the maples intensifies the beauty of the forest in October. An uncommon plant along the Mad River is the Port Orford cedar, found only near the coast as far north as Coos Bay, Oregon, and also in the vicinity of Mt. Shasta, California.

One interesting short hike we enjoyed was the trail to Gray Falls from the adjacent campground. A more rugged but rewarding scenic hike is the trail from near Ruth Reservoir along Devil's Backbone to massive Mad River Rock. The rock can also be reached by a scenic road from the Mad River Ranger Station.

Stanislaus National Forest

Location: 899,000 acres in central California, east of Sonora. State Routes 4, 108, and 120 serve the area. Maps and brochures available at the district ranger stations in Arnold, Groveland, Miwok Village, and Summit or by writing the Forest Supervisor, 19777 Greenley Road, Sonora, CA 95370.

Facilities and Services: Campgrounds (some with camper disposal service); picnic areas with grills; boat ramps; swimming beaches.

Special Attractions: Mokelumne Wilderness; Emigrant Wilderness; reconstructed Miwok Indian Village.

Activities: Camping, picnicking, hiking, backpacking, hunting, fishing, boating, swimming, horseback riding, nature study, winter sports.

Wildlife: Black bear, mountain lion.

SOME OF THE BEST scenic experiences in the Stanislaus are in its Mokelumne and Emigrant wildernesses. We suggest starting your visit in the Mokelumne Wilderness. It is located mostly in the Eldorado National Forest, but it contains a dazzling gorge along the North Fork of the Mokelumne River. At the point just before the river becomes a part of the Salt Springs Reservoir, you will discover Blue Hole, a sparkling pool of blue water that boasts of some of the best fishing in the forest.

The other wild area, the Emigrant Wilderness, lies against the northwest corner of Yosemite National Park. This wilderness has historic significance since it is one of the places through which the gold seekers sought their way across the Sierra Nevadas. The old Emigrant Pass, at the eastern edge of the wilderness, was crossed by 485 people, 135 wagons, and more than 3,000 head of livestock between the years 1851 and 1854. When you visit the pass today, you will wonder how the pioneers were able to blaze their way through. At Saucer Meadow is the grave of one of the emigrants. Other nearby places recalling the era are Emigrant Basin, Emigrant Meadow, and Starvation Lake. There are eight peaks in the wilderness above 10,000 feet, including the highest, Leavitt Peak, at the northeast corner, and the most impressive, Granite Dome, in the center.

Indians roamed this forest land for many decades before the gold seekers. As late as the 19th century, the Miwok Indians occupied the central Sierra Nevadas, living in separate villages and hunting for their livelihood. With the rush for gold through their territory, the Miwoks were driven out, although some of their descendants have returned to live in the area. Near the Summit Ranger Station at Pinecrest, you can visit a reconstructed Miwok village. It is one of the special attractions in the forest. A guide booklet explains life in the village and some of the plants utilized.

Many unusual rock formations are in the Stanislaus. The Dardanelles, an assemblage of interesting rocks north of the Stanislaus River, can be viewed from the Dardanelle Overlook along State Route 108. For a close-up look at strange formations of hardened mud flows, glacial boulders, solidified ash, and iron-filled columns, take the Trail of the Gargoyles which begins six miles up the Herring Creek Road from Route 108. A guide leaflet explains these phenomena, as well as giving them such fanciful names as Devil's Horns,

Gargoyle Ridge, Strip of Fire, River of Stone, Satan's Slipper, Pages of the Earth, and the Wall of Noses.

Several deep and scenic canyons cut into the western slopes of the Sierra Nevadas in the Stanislaus. These have been carved by the Merced, Tuolumne, Mokelumne, Clavey, and Stanislaus rivers. Hiking in any of these scenic areas can also be rewarding.

Tahoe National Forest

Location: 813,181 acres in north-central California, north and west of Lake Tahoe. Interstate 80 and State Routes 20, 49, and 89 are the major roads. Maps and brochures available at the district ranger stations in Downieville, Foresthill, Nevada City, Sierraville, and Truckee, at the Lake Tahoe Visitor's Center, or by writing the Forest Supervisor, Highway 49, Nevada City, CA 95959.

Facilities and Services: Campgrounds (some with camper disposal service); picnic areas with grills; boat ramps; swimming beaches; visitor's center.

Special Attractions: Northernmost stand of giant sequoias; Pacific Crest Trail.

Activities: Camping, picnicking, hiking, backpacking, hunting, fishing, boating, swimming, rock hounding, horseback riding, nature study, winter sports.

Wildlife: Endangered or threatened bald eagle, American peregrine falcon, and Lahontan cutthroat trout; black bear, mountain lion, bobcat, wolverine, golden eagle.

THE TAHOE NATIONAL FOREST is in a beautiful part of the central Sierra Nevadas. The western slopes of this range are dissected by deep, scenic river canyons. The eastern part of the forest is more gently rolling. Elevations range from about 1,500 feet along some of the rivers to 9,400 feet at the Sierra Crest. The forests at the higher elevations are a mixture of red fir, white fir, Jeffrey pine, lodgepole pine, western white pine, whitebark pine, and mountain hemlock. Dropping down to about 5,500 feet, the woods change to a mixture of conifers and hardwoods. The conifers at this elevation are usually sugar pine, ponderosa pine, Douglas fir, white fir, and incense cedar, while the hardwoods include big-leaf maple, California laurel, dogwood, madrone, and black oak. In the foothills of the mountains, usually between 2,000 and 4,000 feet, is a combination of canyon live oak, interior live oak, occasional conifer, and grass. Western junipers grow in some of the drier areas. Along the streams are cottonwoods, willows, and alders.

To get acquainted with plant life in the Tahoe we suggest you start with one of the forest's prime drawing cards — its giant sequoia trees. This is the northernmost location in the world for the giant sequoia. There are only six mature trees in this grove, but they are very special because they are 50 miles north of the next nearest stand. To reach these isolated giants, follow the Auburn-Foresthill road along the Middle Fork of the American River for about 40 miles.

For another wonderful experience, we recommend taking the four-mile-long Rock Creek Nature Trail which begins at the Rock Creek picnic grounds. Among some of the botanical treats you will see are big-leaf maple, white alder, California yew, incense cedar, California hazelnut, madrone, Douglas and white firs, ponderosa and sugar pines, and the green-leaf manzanita.

For another forest excursion, we heartily favor a part of the famous Pacific Crest Trail. This trail, when completed, will extend from Canada to Mexico, a distance of about 2,500 miles. Ninety of these miles are in the Tahoe. Many abandoned mines dating to the gold rush days are along this trail, as well as crystal clear lakes, ridges, and rugged peaks. Since this is gold country, gold particles can be found in the streams. At some of the campgrounds, such as Oregon Creek, Fiddle Creek, and several along the North Yuba River, one of the popular activities is panning for gold. Do not expect to make your fortune this way, however!

INDEX

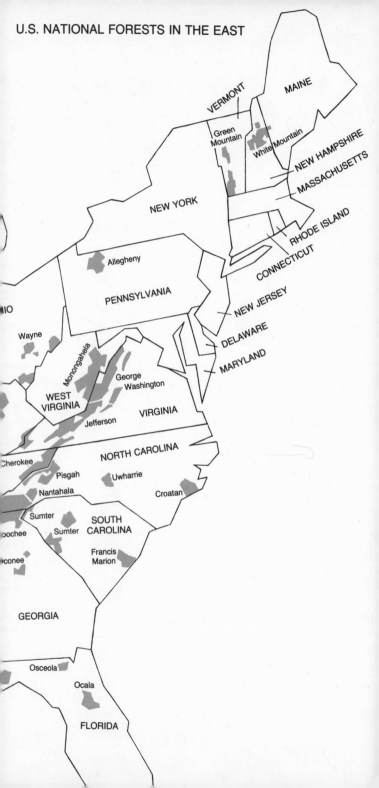

U.S. NATIONAL FORESTS IN THE EAST

VERMONT

MAINE

Green Mountain

White Mountain

NEW HAMPSHIRE

MASSACHUSETTS

NEW YORK

RHODE ISLAND

CONNECTICUT

Allegheny

PENNSYLVANIA

NEW JERSEY

DELAWARE

MARYLAND

OHIO

Wayne

Monongahela

George Washington

WEST VIRGINIA

VIRGINIA

Jefferson

NORTH CAROLINA

Cherokee

Pisgah

Uwharrie

Nantahala

Croatan

Sumter

SOUTH CAROLINA

oochee

Sumter

conee

Francis Marion

GEORGIA

Osceola

Ocala

FLORIDA